Palgrave Studies in the History of Science and Technology

Designed to bridge the gap between the history of science and the history of technology, this series publishes the best new work by promising and accomplished authors in both areas. In particular, it offers historical perspectives on issues of current and ongoing concern, provides international and global perspectives on scientific issues, and encourages productive communication between historians and practicing scientists.

More information about this series at
http://www.palgrave.com/gp/series/14581

Bárbara K. Silva

Astronomy at the Turn of the Twentieth Century in Chile and the United States

Chasing Southern Stars, 1903–1929

palgrave
macmillan

Bárbara K. Silva
Pontificia Universidad Católica de Chile
Santiago, Chile

Palgrave Studies in the History of Science and Technology
ISBN 978-3-030-17711-9 ISBN 978-3-030-17712-6 (eBook)
https://doi.org/10.1007/978-3-030-17712-6

Cover illustration: SOTK2011/Alamy Stock Photo

This Palgrave Pivot imprint is published by the registered company Springer Nature Switzerland AG
The registered company address is: Gewerbestrasse 11, 6330 Cham, Switzerland

Preface

This is the story of the Mills Expedition from the Lick Observatory, California, to Chile, at the beginning of the twentieth century. This expedition built an observatory in Santiago, which is still standing. American astronomers were supposed to stay three years in the southern country, but stayed for twenty-six, time in which they closely related to Chilean society, while trying to unveil some mysteries of the galaxy.

This book is the product of a research process promoted by the Astrophysics Institute of the Catholic University of Chile, which is currently in charge of the Manuel Foster Observatory—formerly known as the Mills Observatory or the Southern Station of the Lick Observatory. Their need to know more about the history of this astronomical place—now an officially designated heritage site—led to this interdisciplinary project. The research aimed to reconstruct the story and explain how an observatory came to the San Cristóbal Hill in the early twentieth century. Historians shed light on the genesis of the observatory while astronomers speak on the tasks and minds of the early scientists.

Research entailed identifying and finding sources in archives in Chile and in California, such as the Press Archive at the National Library of Chile; the National Archive of Administration; the Historical Archive of the Catholic University of Chile; the Mary Lea Shane Archives of the Lick Observatory Endowment at the University of California, Santa Cruz; and the California State Library. Through the sources in these archives—collection of letters, scientific papers, press articles, among many others—we reconstructed the encounters and disagreements between American astronomers and Chileans at the turn of the century. At the same time, these sources allowed us to understand the complex intersection between political, cultural, scientific, and global issues. Combined with the understanding of distinct lifestyles and cultures, the scientific experience opened new approaches to the construction and circulation of scientific knowledge in worlds both strange and familiar.

The study of astronomy reveals how human beings faced the uncertainties of the universe—and how they observed themselves while they transformed knowledge of the stars. At the onset of this research, we learned that the mirrors brought by the astronomers were the key pieces of their equipment. However, and as a metaphor, these were not the only mirrors. The astronomers also brought with them reflections of what occurred across the Pacific. Chile and California were similar geographically and geologically—and it was not just territory: those mirrors were also in the eyes of the protagonists who reflected visions of the "other." Reflections show multiple ways to connect with an American past: Latin American history is intertwined with that of the United States and these entanglements allow us to develop a cultural perspective on the history of science. The reflections also account for urban changes, as well as that southern society that was the host of science. Such an approach shows how politics are inserted, albeit not explicitly, in scientific projections.

Because of this perspective, in this story, science and astronomical progress are told by the protagonists' own words as they systematized their scientific queries, to grasp the cultural background in the early twentieth century astrophysics development. It was a time in which astronomers were trying to understand the galaxy's structure, including the place of our solar system. To study this, scientists needed to observe and analyze the entire sky, and they knew a big portion was not visible from the northern hemisphere: that is the basic reason for the astronomers to travel south.

This book tells the story of the astronomers' adventures in Chile. Chapter 1 is the introduction to the book, which situates the development of astronomy in Chile before the arrival of the Mills Expedition. Then, Chapter 2 focuses on the preparation of the expedition. Obtaining funding for science was a continual challenge, but these preparations also involved the construction of a sophisticated team and the first contacts with Chilean society. Chapter 3 tells of the first experiences of American astronomers in Chile and many of the expressions of the cultural gap.

A labor strike in the port of Valparaíso, as well as the search for a suitable place to erect the observatory, are part of that arrival. Relations with Santiago society, embedded in an era of southern *belle époque*, imbricated with the scientists' adventure in Chile. Soon, they would discover even greater reflections throughout the Pacific, when experiencing the 1906 earthquake.

After a few years in the southern country, they were familiar with some aspects of daily life, but at the same time, they were still constantly surprised—for better or for worse—with the odd habits of Chileans. Chapter 4 addresses the astronomers' adaptation to Chile as they attempted to solve the 'sidereal problem' that had motivated the scientific adventure. They began to gain worldwide recognition and explored territories with a sense of infinity. Their adventures took them to the Atacama Desert, which decades later occupied a fundamental place in global astronomy. Chapter 5 deals with the decades of 1910 and 1920, years of change for the American expedition, after the death of their benefactor and the outbreak of the Great War. It also addresses new relations with the University of California and the final decision to sell the facilities. In this way, the southern adventure of the scientists ended in 1929.

Finally, Chapter 6 is the Epilogue of the book, which tells how this experience of the early twentieth century intertwined with the subsequent development of transnational astronomy in Chile, with a strong presence of American science. There is no linear trajectory between the Lick Observatory expedition and the scientific development of the 1960s; to explain the latter, the years of the Cold War were decisive. The process will culminate in the following years when approximately 70% of the global observation surface will be in Chile.

Perhaps, the global relevance of the Chilean desert for astronomical observation grew from the early adventures of the American scientists from California. Like the mirrors of the spectrograph multiplying the reflections of science, the history of the Southern Station of the Lick Observatory multiplies its interest and its attractiveness. The histories of gringos, Chilenos, scientists, politicians, modernity, science, traditions, and heritage at once reflected and are reflected by the history of a small observatory in the southern hemisphere.

Santiago, Chile — Bárbara K. Silva

Acknowledgements

Historical research takes quite some time: designing the research, never-ending hours in archives, processing all the information and of course long periods of writing. For those who chose working in the humanities, we learn to live with these timelines, and research along with the writing processes become part of our lives. That is why I need to say thanks to all those people who were with me when I decided it was time for a change, and got into history of science. Thanks to all of you I submerged myself in the history of a tiny—kind of hidden—astronomical observatory in Santiago, and was able to give new meaning to mi previous years working in cultural and political global history. Some historians are suspicious about getting into new areas of research; I am very thankful I am not one of those historians, and that I could give this charming southern observatory a new life, the one it gained through these pages.

The story of this book began in 2015, and would have never been possible without the eagerness of Olaya Sanfuentes and Gaspar Galaz. Thanks to them I got involved in history of astronomy, to rebuild the history of this observatory, at the summit of a landmark hill in Santiago de Chile. We decided to apply for a small grant—an Interdisciplinary Project, of the Research Vice-Rectory of the Pontificia Universidad Católica de Chile, shared by the Astrophysics and the History Departments. Back then, I was still struggling to finish my Ph.D. Dissertation, and to start a new research seemed dangerous yet so attractive. I am so glad I took the chance. At first, I could not even understand the words I found on the Astronomers' documents. Luckily, I decided to get involved into that problem, which also allowed me to learn and value working with other experts. I am deeply thankful to Gaspar and Olaya, for every story needs a beginning, and I owe this to them.

After one year, the funding provided by the University was over, but I decided to continue working in this amazing adventure of science at the turn of the century. The continuous support of the Astrophysics Department was crucial to this book. In this process, Leonardo Vanzi, from the Astro-Engineering Center was crucial. The interest and persistence of Leo Vanzi deserve special thanks. He was permanently involved in our progress, and through some of his projects, we could to go to California to try to find if there were any sources regarding this story of the Lick Observatory adventure in Chile. There were so many!

In California, I am very thankful of the patience, help and kind support of Luisa Haddad at the Mary Lea Shane Archives, in Special Collections of University of California Santa Cruz. It is a privilege for a historian to find people like Luisa in charge of an archive. In the same way, I thank Paul Lynam, astronomer at the Lick Observatory, who let my visit there to be a great experience.

Of course, it is my pleasure to thank our research assistants, students who were brave enough to take part in this unusual project. Special thanks to Bárbara Ossa and Francisca Espinosa; they were amazing in retrieving bibliography and archival work in Chile. Diego Caro was a great assistant, who was able to manage and organize an enormous amount of sources, and also was a partner to discuss the structure of the book. I thank Nicolás López, who has worked with me for so many years now, and I was lucky enough he got involved in the manuscript, and helped me in digging for some last minute sources. I am very grateful of all of you, especially for the kind and vibrant energy you gave to this project. Afterwards, when I finished the manuscript Alejandro Clocchiatti, professor at the Astrophysics Department, patiently revised it for scientific accuracy. Thank you so much for this, Alejandro.

I never thought of writing a book in English, and when the possibility came up, the Chile California Council made it possible. Their enthusiasm about the book and this new perspective on the Chile-California relationship led them to finance the editing of the manuscript. Robert Weis, a fantastic bilingual historian, worked on the entire draft, helping me with enormous patience to convert my rough English into smooth reading. Robbie, it is impossible to think of a better editor/translator than you. The FONDECYT project 3170099–Chile allowed me to enhance my knowledge in history of astronomy, and to have the time to write this manuscript. I also want to thank Palgrave Macmillan for trusting this book and me, and the reviewers and their thoughtful suggestions.

Last but not least, I want to thank friends and colleagues who supported me in this insane and lovely project I took between a Ph.D. dissertation, other publications, and the every day work at the university. I want to give special thanks to William San Martin and Alvaro Rodriguez, who helped and accompanied me in my research in California, always cheering me up. So many friends, historians and non-historians, supported me with love and enthusiasm, even when it seemed I was doing something completely crazy: Alfredo Riquelme, Rafael Gaune, María Montt, Fernando Purcell, Claudio Rolle, Carla Rivera, Camila Gatica, Kamyar Aram, Abhijeet Kulkarni, Vero Aleotti, Cassandra Young, and my partners at the History of Science and Technology Laboratory.

All of you are part of this adventure. "Stellar thanks" to each one of you.

Contents

CHAPTER 1

Introduction: Look Up to Southern Stars

Abstract The introduction opens with remarks on how astronomy connects us with our past; every human being, in every time or place, has done the same gesture: looking up, to those lights within the darkness. The introduction presents the development of astronomy in Chile before the Mills Expedition arrival. There was only one previous initiative in the mid nineteenth century, also led by Americans—in this case, by Lt. James Melville Gilliss. Afterwards, I present the background of James Lick, the donor of the Lick Observatory. Lick's life takes us to South America in the 1830s, in his longing to become wealthy and get the love of his life.

Keywords Astronomy in Chile · James Lick · Nineteenth century

After two months at sea, the American astronomer, William H. Wright finally arrived in the port of Valparaíso on April 18, 1903. His wife and his assistant H. K Palmer were traveling with him. He had shipped out of San Francisco on February 28 to lead an astronomical mission. The mission was astronomical indeed—not only due to nature of the study but also to dimensions of the task. The team of astronomers was about to begin research on stars of the southern hemisphere. The goal entailed traveling from one end of the planet to the other, with all the necessary equipment. To cross the ocean, with an enormous telescope and dozens of delicate pieces of observatory equipment, they chose the steamship *Perú*.

B. K. Silva, *Astronomy at the Turn of the Twentieth Century in Chile and the United States*, Palgrave Studies in the History of Science and Technology, https://doi.org/10.1007/978-3-030-17712-6_1

The research infrastructure was modeled on the Lick Observatory in Mount Hamilton, near San Jose California. The new observatory was not going to be identical, but the team talked about the "southern twin"—a similar observatory in the faraway southern latitudes. The ship brought the dome that would cover the cutting-edge technology and thus allow them to study the stars and deepen their knowledge of the universe. They also shipped a spectrograph and the complex puzzle of pieces that made it work. The pieces included several precise mirrors that generated the reflections needed for spectroscopy and constituted a window to the newly opened field of astrophysics. By registering the light or spectra of brighter stars, they derive an understanding about stars that were invisible from the mountains in California.

When Wright gazed at Valparaíso, he had a sense that the mirrors were also talking about other dimensions. They too had traveled thousands of miles southward and now faced scenery that was at once exotic and oddly familiar. The landscapes of Chile resembled their home near the Sacramento River delta, and the vegetation, the weather, and the mountains they saw from the ocean provoked a strangely recognizable feeling.

The long years of preparation for the mission ended here in the south. Wright began a new phase in his scientific and human experience. It was going to be a brief phase, Wright thought. He had yet to discover that their work in the southern hemisphere was going to take much longer than they had imagined. He could not have foreseen the importance that astronomy would take on, decades later, in the tiny country at the end of the world. Nor could he imagine that his work would become elemental to the development of science in Chile. He was just a gringo astronomer trying to get off a ship in a southern Pacific port, where he would commence a personal adventure toward unknown spheres of science and transnational relationships.

Thus begins the story of how a little observatory was built in Santiago, Chile on the summit of San Cristóbal Hill. More than a century later, in 2010, the San Cristóbal Observatory was named a historical monument in a country that had become a world astronomical center.[1] More than a century earlier, American scientists launched one of Chile's first astronomical adventures. This is that story.

[1] Consejo de Monumentos Nacionales de Chile, *Decree number 0352*, September 3, 2010.

1.1 Astronomy in Chile Before the Mills Expedition

Understanding the universe has always been a dimension of human inquisitiveness. Every civilization has looked to the skies, wondering what is "up there" where human eyes can no longer see. The possibility of understanding the stellar structure has the most diverse consequences. It is about scientific discovery, certainly; yet, it also rooted in the comprehension of humanity and in the changing visions of the world.[2] Trying to learn about the stars, astronomers had a global dimension that inevitably brought together different cultures and societies.[3] From the ancient Greeks to the American indigenous cultures, looking to the skies has been part of the human dimension of knowledge.[4] Afterwards, the creation and spread of medieval European universities integrated observation and study of the stars.[5] During the era of European discoveries and explorations, travelers needed navigation techniques that relied on astronomy. With pioneers such as Copernicus, Brahe, Kepler, and Galileo, astronomy left a profound mark on modern Europe's scientific production.[6]

Chile was no exception to this stellar curiosity. The Spaniards who arrived at these lands developed the first written records of southern stars.[7] Alonso de Ovalle's *Historica Relacion del Reyno de Chile* features a whole chapter on the stars and constellations.[8] The Spanish

[2]Thomas Kuhn, *The Structure of Scientific Revolutions* (Chicago: The University of Chicago Press, 2012 [1962]), 111.

[3]Lissa Roberts, "Situating Science in Global History: Local Exchanges and Networks of Circulation," *Itinerario* 33, no. 1 (2009): 9–30.

[4]Jesus Galindo, *Arqueoastronomía en la América Antigua* (Madrid: Sirius, 1994).

[5]Edward Grant, *Physical Science in the Middle Ages* (Cambridge: Cambridge University Press, 1977), 60ff; Jacques Le Goff, *Los intelectuales en la Edad Media* (Barcelona: Gedisa, 1990), 139–149.

[6]René Taton and Curtis Wilson, eds., *Planetary Astronomy from the Renaissance to the Rise of Astrophysics* (Cambridge: Cambridge University Press, 1989).

[7]With this statement, the intention is to recall initiatives related to modern science, and not those of ancestral concerns and knowledge regarding the stars. Indigenous people of Chile certainly studied stellar movements. Although fascinating, that is part of another story and I do not include it in this book.

[8]Alonso de Ovalle, *Historica relacion del Reyno de Chile y de las missiones y ministerios que exercita en la Compañía de Jesus* (Roma: Francisco Cavallo, 1646).

crown instructed explorers to observe a lunar eclipse in 1582, which became the first record of astronomical observation in Chilean territory.[9] Centuries later, in the aftermath of independence, Bernardo O'Higgins intended to build an astronomical observatory in Chile; unfortunately, it never materialized.[10] Although these observations and project were never more than ephemeral, they shed light on the fascination in the stars that reached the end of the world.

The astronomical team led by Wright in the early twentieth century was not the first in Chile. A different American mission, led by Lt. James Melville Gilliss, arrived in 1849. A year before, following a proposal from the United States Naval Observatory, the United States Congress ordered the formation of a scientific mission that was to travel to the southern hemisphere in order to observe the movements of Venus and Mars and thus measure the distance between the sun and the Earth through solar parallax.[11] Gilliss arrived in Chile with three assistants and all the equipment needed to install a provisional observatory on the summit of Santa Lucía Hill in downtown Santiago. At the time, astronomical science was linked to other disciplines such as seismology or atmospheric sciences. According to the nineteenth-century Chilean historian, Diego Barros Arana, the goal of the mission was the following:

> [They] had to observe, simultaneously with the northern observatories, Venus and Mars during their respective stationary terms and oppositions in the years 1849, 1850 and 1852. The commission would also make other astronomical observations in the southern hemisphere, and also those regarding meteorology and magnetism, and those which would be possible to make regarding the earth's tremors.[12]

[9] Hernán Quintana and Augusto Salinas, "Cuatro siglos de astronomía en Chile," *Revista Universitaria* 83 (2004): 54.

[10] Ricardo Leiva, "Atacama: a 100 años del informe Curtis," *Revista de Humanidades*, 3 (2010): 12.

[11] Sebastián López et al., "Astronomía," *Análisis y Proyecciones de la Ciencia Chilena* (Santiago: CONICYT, 2005), 240; Patricio Rogers, "La Astronomía en Chile durante la segunda mitad del siglo XX," *Revista Chilena de Historia y Geografía* 150 (1982): 30.

[12] Diego Barros Arana, *Un decenio de la historia de Chile 1841–1851* (Santiago: Universitaria, 1905–1906), 405.

The expedition was a novelty but its operations were limited and modest. Even so, it was more successful than the previous attempt by the Chilean government. Officials had discussed building an observatory but never carried through, due, according to the Chilean minister in Washington, Manuel Carvallo, to the lack of men with appropriate scientific knowledge.[13] Perhaps this unfulfilled desire explains the warm welcome the government gave to Gilliss.

The hospitality went beyond protocol. When the Americans decided to build the provisional observatory on Santa Lucía, local officials leveled part of the hill's summit. In exchange, the Chilean government asked for compensation that would help fund a national observatory at some time in the future. Furthermore, officials requested that the Americans sell them their equipment upon the completion of the mission. Three Chileans were to join the Gilliss Mission in order to learn the techniques required to handle the instruments and make the corresponding observations.[14] The goal was to acquire the technology and technical expertise necessary to develop the National Astronomical Observatory.[15] Scholars have argued that astronomy in Chile began "as a direct consequence" of the Gilliss mission, in the mid-nineteenth century.[16]

Gilliss spent three years in the country working on his scientific tasks. Gilliss and his team members built solid relations with Chileans and their institutions, especially with the University of Chile, which appointed him as an honorary member of the Faculty of Physics and Mathematics.[17] Despite the Chilean government's efforts to convince Gilliss to stay and oversee the country's astronomy program, the lieutenant returned to the United States in 1852.[18] Back in the United States, Gilliss presented

[13] Barros Arana, *Un decenio,* 406.

[14] Arturo Aldunate Phillips, *Los grandes observatorios astronómicos instalados en Chile* (unpublished, 1977), 10.

[15] Augusto Salinas, *La ciencia en Chile y en los Estados Unidos de Norteamérica: un análisis histórico comparado (1776–1976)* (Santiago: CONICYT, 1976), 119.

[16] Rogers, "La Astronomía en Chile", 29.

[17] Salinas, *La ciencia en Chile,* 119.

[18] In fact, afterwards Gilliss was appointed in charge of the United States Naval Observatory, where he experienced the civil war. Steven J. Dick, *Sky and Ocean Joined. The U.S. Naval Observatory 1830–2000* (Cambridge: Cambridge University Press, 2003), 140ff.

four massive volumes to the American Congress. He described not only his astronomical discoveries, but also his observations on topics as varied as fauna and flora, the urban landscape of Santiago, meteorology, and seismology.[19] Gilliss helped Chile acquire Smithsonian publications and, in return, the Chilean government sent a collection of the *Annals of the University of Chile* to the Smithsonian Institution. Thus began an exchange of magazines and books between the two institutions that continued through the nineteenth century.[20]

With his departure imminent, the government was able to purchase many of the mission's instruments.[21] Karl Wilhelm Moesta, one of Gilliss's local assistants, remained in charge of the Chilean government's new project.[22] A four-inch meridian circle and a six-inch equatorial refractor formed the foundations of the National Observatory, associated with the recently founded University of Chile. The National Astronomical Observatory was born, in the middle of the nineteenth century, on the summit of Santa Lucía Hill. Around 1856, the Observatory moved to Quinta Normal, because of the instrumental instability due to the oscillations detected in the hill, by then attributed to the daily movement of the Sun. The new facilities were completed in 1862.[23]

1.2 The Philanthropist's Love: James Lick

These experiences with travel, exchange, and relationships between Americans and Chileans occurred again in the early twentieth century. The new expedition was linked to James Lick, who, quite apart from scientific endeavors, had his own adventure in the southern hemisphere. Indeed, the reflections between Chile and California connected much more than similar vegetation and wildlife. Mirrors reflected not only

[19] James M. Gilliss, *The US Naval Expedition to the Southern Hemisphere* (Washington, DC: A.O.P. Nicholson Printer, 1855); Hilmar W. Duerbeck, "National and International Astronomical Activities in Chile," *Astronomical Society of the Pacific Conference Series* 292 (2003): 4.

[20] Salinas, *La ciencia en Chile*, 118–119.

[21] Barros Arana, *Un decenio*, 408.

[22] Duerbeck, "National and International Astronomical," 5.

[23] Leiva "Atacama: a 100 años," 14.

geographies, but also human experiences. Events in the United States during the 1820s were about to connect with scientific developments in Chile in unsuspected ways.

James Lick, a young carpenters' son in Pennsylvania had fallen in love with Barbara Snavely. Shortly after, Snavely discovered she was pregnant. Lick approached her father to ask for her hand in marriage. But Henry Snavely decided that Lick was not good enough for his daughter. He told the young man that he could only marry Barbara once he had land that was as big and valuable as his own.[24] Heartbroken and disappointed, Lick resolved to build a fortune, face Henry Snavely, and marry Barbara. He left his hometown for Baltimore, where he learned to build and tune pianos. His progress was enormous. In just a few months, he had his own piano business in New York. Now, he wanted to further accelerate the acquisition of his fortune. He learned that the piano business was highly regarded and better paid in South America so he moved to Buenos Aires, Argentina.[25] In 1832, considering his fortune sufficient to ask for Barbara's hand, Lick returned to Pennsylvania only to discover that Barbara had married another man and had left town with their son.[26]

Miserable, Lick went back to South America, his motivation to resettle in the United States having vanished. With no purpose left in Argentina, Lick moved on to Valparaíso, Chile. Though a medium-sized Pacific port, as the mandatory stop after the Straits of Magellan, Valparaíso was crucial for interoceanic navigation.[27] An appealing place for immigrants, it gradually grew into a cosmopolitan city. After a few years, Lick left Chile to try his luck in the Peruvian market. He settled in Lima in 1836, where his fortune so quickly grew that he remained for almost a decade.

[24] Rosemary Lick, *The Generous Miser: The Story of James Lick of California* (Los Angeles: The Ward Ritchie Press, 1967).

[25] Donald E. Osterbrock, John R. Gustafson, and Shiloh Unruh, *Eye on the Sky: Lick Observatory First Century* (Oakland: University of California Press, 2010), 3. Authors detail a series of adventures Lick experienced when going back to Argentina from a trip to Europe, including nearly sinking in the ocean and the escape from the Brazilian army.

[26] Lick, *The Generous Miser.*

[27] Bárbara Silva y Josefina Cabrera, *Chile, 100 días en la historia del siglo XX* (Santiago: Planeta, 2015), 57ff.

New opportunities appeared with conflicts between the United States and Mexico over the independence and annexation of Texas. The disagreements between the White House and the unstable Mexican government increased and the two countries declared war. The American victory was overwhelming. Amid administrative and governmental crises, Mexico signed the agreement that gave the United States most of the territories in dispute, including California.[28] Soon after, in early 1848, James Lick left Peru for San Francisco. The American far west was a place of freedom, where it was possible to start a new life, beyond lineages, surnames, ancestry, and traditions. For the Americans, it was a new land.

Through the piano business, Lick had already built a substantive fortune, so money was not a problem. The decision to invest his fortune in property seemed absurdly risky at the time, but Lick began to buy lands in the Bay Area as soon as he arrived. Then, the Gold Rush changed San Francisco and its surroundings drastically. With hundreds of people dreaming of gold streaming in, the quiet town quickly became a bustling city.[29] Many came from the East coast and the trip to California was difficult. One of the most viable options was to go around South America, through the Straits of Magellan, and stop in Valparaíso before continuing toward the Northern Pacific.[30] The routes that sought to be efficient and safe linked California and the southern Pacific coast. These connections through the Pacific are fundamental to understanding the arrival of American astronomers in Chile a half a century later.

Now a magnate of the West, Lick remained a lonely man. His motivation to avenge himself for Henry Snavely's snub no longer drove him. The love of his youth, Barbara, was dead. When his son John was 37, he took him to California, but their relationship was another source of disappointment. John returned to the East Coast and Lick left him

[28] Andrés Reséndez, "Guerra e identidad nacional," *Historia Mexicana* 47, no. 2 (1997): 412–413

[29] Fernando Purcell, *¡Muchos extranjeros para mi gusto! Mexicanos, chilenos e irlandeses en la construcción de California, 1848–1880* (Santiago: FCE, 2016), 51.

[30] H.W. Brands, *The Age of Gold: The California Gold Rush and the New American Dream* (New York: Anchor, 2003), 103–121.

out of his will.[31] Years went by. Lick grew old, lonesome, and with a fortune beyond his dreams. He had built an enormous luxury hotel in San Francisco but he yearned to remain in California's memory in other ways. Sick, his musings over what kind of legacy he would leave were his greatest concerns. He did have an idea, probably suggested by George Madeira,[32] to build a giant telescope. This project was consigned and detailed in his deed of trust. Lick set aside $700,000—quite a fortune at the time—"for the purpose of purchasing land and constructing and putting up on such land a powerful telescope, superior to and more powerful than any telescope ever yet made."[33] The University of California was to take charge of these facilities. This would be his legacy and his memory. The telescope and the observatory would be called the Lick Astronomical Department of the University of California.

Before the telescope was completed, Lick died. His fortune never bent the will of the Snavely family but it did finance the construction of the observatory in Mount Hamilton near San Jose, California. During construction, Lick's grave was moved to the observatory's basement,[34] his remains laying below the most powerful telescope in the world, watching and taking care of the astronomers who labored on that summit overlooking the San Francisco Bay. Twenty years after Lick's death, one of those astronomers, William Wallace Campbell, began to study the possibility of installing an observatory in the southern hemisphere. Lick never imagined that astronomers would visit and work in the same places he did in his own journey of love and disappointment to South America. Now the motivation was science and those who ventured south were the Lick Observatory astronomers.

[31] Lick, *The Generous Miser.*

[32] Donald E. Osterbrock, *James E. Keeler: Pioneer American Astrophysicist and the Early Development of American Astrophysics* (New York: Cambridge University Press, 1984), 38.

[33] *Deed of trust of James Lick*, dated September 21, 1875, and recorded in the Office of the Recorder, of the County of San Francisco, State of California, November 10, 1875. B1464518x, San Francisco Public Library, California, 17.

[34] Lick grave is actually under one of the Lick Observatory's domes. With an amazing system of gears and pulleys, the ground moved up and down the base of the telescope, which allowed a more precise observation. On this way, there was a space below the dome, some sort of basement, where Lick's remains still lie, under the telescope he never got to see.

Bibliography

Aldunate Phillips, Arturo. *Los grandes observatorios astronómicos instalados en Chile*. Unpublished, 1977.

Barros Arana, Diego. *Un decenio de la historia de Chile 1841–1851*. Santiago: Universitaria, 1905–1906.

Brands, H.W. *The Age of Gold: The California Gold Rush and the New American Dream*. New York: Anchor, 2003.

Consejo de Monumentos Nacionales de Chile, *Decree Number 0352*, September 3, 2010.

Deed of trust of James Lick, dated September 21, 1875, and Recorded in the Office of the Recorder, of the County of San Francisco, State of California, November 10, 1875. B1464518x, San Francisco Public Library, California, 17.

Dick, Steven J. *Sky and Ocean Joined: The U.S. Naval Observatory 1830–2000*. Cambridge: Cambridge University Press, 2003.

Duerbeck, Hilmar W. "National and International Astronomical Activities in Chile," *Astronomical Society of the Pacific Conference Series* 292 (2003): 3–20.

Galindo, Jesus. *Arqueoastronomía en la América Antigua*. Madrid: Sirius, 1994.

Gilliss, James M. *The US Naval Expedition to the Southern Hemisphere*. Washington DC: A.O.P. Nicholson Printer, 1855.

Grant, Edward. *Physical Science in the Middle Ages*. Cambridge: Cambridge University Press, 1977.

Kuhn, Thomas. *The Structure of Scientific Revolutions*. Chicago: University of Chicago Press, 2012 [1962].

Le Goff, Jacques. *Los intelectuales en la Edad Media*. Barcelona: Gedisa, 1990.

Leiva, Ricardo. "Atacama: a 100 años del informe Curtis," *Revista de Humanidades* 3 (2010): 10–22.

Lick, Rosemary. *The Generous Miser: The Story of James Lick of California*. Los Angeles: The Ward Ritchie Press, 1967.

López, Sebastián, Mario Hamuy, Luis Felipe Barrientos, José Maza, and María Teresa Ruiz. "Astronomía." In *Análisis y Proyecciones de la Ciencia Chilena*, 239–260. Santiago: CONICYT, 2005.

Osterbrock, Donald E. *James E. Keeler: Pioneer American Astrophysicist and the Early Development of American Astrophysics*. New York: Cambridge University Press, 1984.

Osterbrock, Donald E., John R. Gustafson, and Shiloh Unruh. *Eye on the Sky: Lick Observatory First Century*. Oakland: University of California Press, 2010.

Ovalle, Alonso de. *Historica relacion del Reyno de Chile y de las missiones y ministerios que exercita en la Compañía de Jesus*. Roma: Francisco Cavallo, 1646.

Purcell, Fernando. *¡Muchos extranjeros para mi gusto! Mexicanos, chilenos e irlandeses en la construcción de California, 1848–1880*. Santiago: FCE, 2016.

Quintana, Hernán and Augusto Salinas, "Cuatro siglos de astronomía en Chile," *Revista Universitaria* 83 (2004): 53–60.

Reséndez, Andrés. "Guerra e identidad nacional," *Historia Mexicana* 47, no. 2 (1997): 411–439.

Roberts, Lissa. "Situating Science in Global History: Local Exchanges and Networks of Circulation," *Itinerario* 33, no. 1 (2009): 9–30.

Rogers, Patricio. "La Astronomía en Chile durante la segunda mitad del siglo XX," *Revista Chilena de Historia y Geografía* 150 (1982): 29–48.

Salinas, Augusto. *La ciencia en Chile y en los Estados Unidos de Norteamérica: un análisis histórico comparado (1776–1976)*. Santiago: CONICYT, 1976.

Silva, Bárbara and Josefina Cabrera. *Chile, 100 días en la historia del siglo XX*. Santiago: Planeta, 2015.

Taton, René and Curtis Wilson, eds. *Planetary Astronomy from the Renaissance to the Rise of Astrophysics*. Cambridge: Cambridge University Press, 1989.

CHAPTER 2

Preparing the Mills Expedition

Abstract This chapter focuses on the preparation of the Expedition, by addressing how the American astronomers organized it, and why they decided to travel to Chile. Preparations involved getting the funding from a wealthy banker from Sacramento (Mr. Mills), but also how they built all the sophisticated equipment they needed. By connecting different enterprises, industries and expertise, in different places of the United States, they managed to have all the pieces in Mount Hamilton. This chapter also describes first negotiations with Chileans, and the story of how their desire for going south took shape, between unexpected problems and personal expectations.

Keywords Mills Spectrograph · Chilean history · Mills Expedition

Astronomy was already a prestigious science worldwide by the mid-nineteenth century. Yet its development was not equally distributed; it was significantly more intense in the northern hemisphere. Struggling to meet basic needs and overcome long histories of colonization and imperial domination, southern countries still grappled with how to embrace the paradigm of progress. A sophisticated science such as astronomy did not top their priorities. Still, the science reached some areas of the southern hemisphere, such as Peru, South Africa, and Australia.

B. K. Silva, *Astronomy at the Turn of the Twentieth Century in Chile and the United States*, Palgrave Studies in the History of Science and Technology, https://doi.org/10.1007/978-3-030-17712-6_2

In the late nineteenth century, astronomers from the Lick Observatory in California were determined to build an observatory somewhere in the southern part of the world. The scientific community insisted "there was an urgent need for systematic determinations of radial velocities of southern stars." Therefore, Lick astronomer William Wallace Campbell began organizing an expedition to install an observatory in the southern hemisphere.[1] Campbell's first attempts to send an expedition to the southern hemisphere in 1894 sought to "complement astronomical observations of the northern hemisphere, on stars' radial velocities."[2] The goal was ambitious but vital. Measuring radial velocities of the brighter stars could help astronomers solve what Campbell called "The Sidereal Problem." He explained it was about understanding the galaxy dynamic and how matter is distributed in the universe, assuming that the position and trajectories of celestial bodies depend on their own movement but also on the observer's point of view.[3]

2.1 Money, Science, and Mr. D.O. Mills

Before Campbell's team could begin negotiating with local authorities, or even decide which southern country to pick as a host, their expedition needed funding. Fortunately, they could turn to the philanthropically inclined social circles in the United States. The Lick Observatory had established a close relationship with Darius Ogden Mills. Like James Lick, Mills settled in Sacramento at the outbreak of the Gold Rush.[4] Mining and trade were expanding in the West Coast, but Mills chose to get involved in finance and railroads investments.[5] Shortly after his first foray

[1] Christiaan Sterken, "Radial Velocity Observations with the 36″ Telescope at Cerro San Cristóbal, Santiago, Chile," *The ESO Messenger*, no. 28 (1982): 12–13.

[2] William W. Campbell, "Organization and History of the D.O. Mills Expedition to the Southern Hemisphere," *D.O. Mills Expedition, Cerro San Cristóbal, Santiago, Chile from Its Organization in December 1900 to March 1906*, ed. W.W. Campbell and W.H. Wright, Vol. IX (Sacramento: W.W. Shannon Superintendent State Printing, 1907), 5.

[3] D.O. Mills Expedition, February 28, 1903, UA 36. Ser 04. Box 8 Folder 1, *Lick Observatory Records*, Special Collections, Mary Lea Shane Archives, University of California Santa Cruz. The Sidereal Problem is explained in full, according to Campbell's words in Chapter 4.

[4] "Darius Ogden Mills," *New York Times*, November 27, 1898.

[5] "D.O. Mills," *National Bank of D.O. Mills and Company Records, 1847–1927*, Manuscript 368–378, California State Library.

into the conquest of the West, Mills funded the Gold Bank of D.O. Mills & Co. in Sacramento and invested in the Bank of California in 1864.[6] By the mid-1870s, he moved his business headquarters to New York; however, his relationship with California persisted. Indeed, he bought many buildings in downtown San Francisco and an extensive property in nearby rural Millbrae. He could trace his ancestors back to America's colonial settlers, but for society, and in his own eyes, he was a self-made man.[7]

Mills's attitude toward technological progress led him to different projects, most of which related to his investments in the railroad business. As a man of status, he also needed to engage in social activities such as philanthropy for science.[8] By the late nineteenth century, society's trust in science was widespread and almost unquestioned. Human beings were expanding their knowledge and the possibilities of understanding themselves and their environment. "The progress of science and invention has increased the opportunities a thousand-fold," Mills remarked. "A man can find them wherever he seeks them, in the United States in particular. It has caused the field of employment of labor of all kinds to expand enormously, thus creating opportunities which never existed before."[9] Progress in fields such as medicine and psychology seemed astonishing. Biology and microbiology broadened the known microscopic universe; botany and zoology, among many other areas of exploration, accelerated and multiplied constantly. Scientific associations helped specialize and differentiate knowledge. New discoveries and inventions moved the horizon of what was possible: humans could do and learn things they had never imagined.

At the same time, ancestral concerns crossed into this climate of scientific novelty. Driven by this sprit, the observation of the stars underwent repeated updates. Long removed from the Renaissance, scientists discovering the universe were newly amazed by the star-filled sky. The origins of mankind were there, somewhere. Mills, the Gold Rush entrepreneur, wanted to be part of the vanguard of science—astronomy. He joined the

[6] "Darius Ogden Mills," *New York Times*, November 27, 1898.

[7] Orison Swett Marden, "Darius Ogden Mills: A Self-Made Man Who Strives to Give Others a Chance," *Little Visits with Great Americans: Or Success, Ideals, and How to Attain Them*, Vol. 1 (New York: The Success Company, 1905), 118.

[8] Barry D. Karl and Stanley N. Katz, "The American Private Philanthropic Foundation and the Public Sphere 1890–1930," *Minerva* 19, no. 2 (June 1981): 236–270.

[9] Swett, "Darius Ogden Mills," 120.

Administrative Board of Lick Observatory.[10] He strengthened his relationship further in 1897, when he donated funds to build the observatory's spectrograph, which was named after him.[11]

Astronomers were beginning to use spectrographs more extensively to study the stars' energy through their light, or spectra. By observing the frequency range, or the visible wavelength of a star's spectrum, astronomers could discover information about its energy. Therefore, spectroscopy has to do with velocities and distribution of the energy of the stars, rather than with the positioning of the stars in the sky. The California scientists focused their spectroscopy on the measurement of stars' radial velocities. That is, they measured how fast a body moved toward or away from their line of sight.[12] Campbell wrote that, "A three-prism spectrograph, built in our instrumentation workshop, based in my designs, resulted in many conferences."[13] Indeed, Campbell and Wright's innovations in spectroscopy were producing promising outcomes.

The Mills Spectrograph was fundamental for advances in astronomy. The relationship between Mills and Lick Observatory grew stronger. When Campbell decided to travel to the southern hemisphere to study radial velocities of southern stars, he turned to Mills. Still, getting the amount money they needed would not be easy. When astronomers from Mount Hamilton approached him in 1900, Mills replied that he was, "rather overburdened with donations of various kinds, and should, therefore, hardly wish to entertain this. Trusting you have found someone to do the needful, and with my best wishes for the University, and the Lick Observatory."[14] Nonetheless, by the end of that same year, Mills changed his mind. He decided to invest an enormous amount in the expedition to the southern hemisphere. In December 1900, Mills wrote Campbell and noted that he was now "inclined to furnish the

[10]D.O. Mills Expedition, Correspondence—D.O. Mills to Edward S. Holden, New York, September 24, 1895, UA 36. Ser 04. Box 8 Folder 2, *Lick Observatory Records*, Special Collections, Mary Lea Shane Archives, University of California Santa Cruz.

[11]William W. Campbell, "A Brief Account of the D.O. Mills Expedition to Chile," *Publications of the Astronomical Society of the Pacific* 15, no. 89 (April 1903): 70–75.

[12]Ian Ridpath, *Dictionary of Astronomy* (Oxford: Oxford University Press, 1997), 441

[13]Campbell, "Organization and History," 6.

[14]D.O. Mills Expedition, Correspondence—D.O. Mills to J.E. Keeler, Director of the Lick Observatory, New York, March 12, 1900. UA 36. Ser 04, Box 8 Folder 2, *Lick Observatory Records*, Special Collections, Mary Lea Shane Archives, University of California Santa Cruz.

money for the expedition to observe the southern skies as set forth by you. If this work can be accomplished in the two years for $24,000.00, I shall take pleasure in supplying the funds."[15]

Campbell's dream was coming true. With Mills's funding, they could plan an astronomical expedition to the Southern Hemisphere, "with purpose to observe the radial velocities of about 175 of the brighter stars whose spectra contain well-defined lines."[16] It was so exciting that years later he would omit all the negotiations that attaining the financing required. The relationship with his benefactor meant he could continue the efforts he had been working toward for years. With Mills and his money on board, the next step was for the scientists to plan their adventure in the southern world.

2.2 The Southern Longing Becomes Real

A few days after Mills's promising response, the expedition's objectives were clear. With the 24,000 Mills had set aside, Campbell could provide for a two-year expedition and continue work on "the problem of the movement of stars in the line of sight." As Campbell noted, "Many valuable results have already been secured, chiefly by the use of the Mills Spectrograph, the finest instrument of its kind in the world and presented to the observatory by D. O. Mills."[17] While praising and recognizing his benefactor, Campbell detailed the purpose of their tasks in the south. Less clear was where to send the expedition. The southern hemisphere, though not entirely unknown, was a strange world. Around 1900, the Lick astronomer still equivocated between South America and Australia.[18] Oceania's cultural proximity to the United Kingdom

[15] D.O. Mills Expedition, Correspondence—D.O. Mills to W.W. Campbell, Acting Director of the Lick Observatory, New York, December 1, 1900. UA 36. Ser 04. Box 8 Folder 2, *Lick Observatory Records*, Special Collections, Mary Lea Shane Archives, University of California Santa Cruz.

[16] William W. Campbell, "Notes from Pacific Coast Observatories," *Publications of the Astronomical Society of the Pacific* 40, no. 236 (1928): 250.

[17] D.O. Mills Expedition—Arrangements, December 9, 1900. UA 36. Ser 04. Box 8 Folder 1, *Lick Observatory Records*, Special Collections, Mary Lea Shane Archives, University of California Santa Cruz.

[18] "Berkeley, December 8.- Through the generosity of D.O. Mills, who has promised to defray the expenses, the University of California will send an astronomical expedition from the Lick Observatory to Australia or South America to study the stars of the southern

conferred a degree of security and the Anglophone population would ease communication for the planned two-year stay.

Although spectroscopy was in full development, without examination of the southern skies, the astronomers' information was incomplete. As Campbell later wrote, he was impressed by the urgent need of observations on radial velocity, especially in the southern hemisphere. He then explained the ignorance regarding the southern skies: "Although we know the movement of the solar system through space or know the structure of the sidereal system, we certainly cannot wait to base a satisfactory solution on radial speeds of the stars confined in a half or two-thirds of the sky."[19] Of course, Campbell knew there were previous experiences in the south: there was Gilliss's trip to Chile and there had been direct observations taken in South Africa and Argentina.[20] Harvard astronomers had carried out photometric observations in Arequipa, Peru and the British had missions in Cape of Good Hope.[21] And there were other experiences in Melbourne and Adelaide, Australia.[22] However, the understanding of the universe and its movement required a more solid scientific foundation, one that could only come from extensive studies from the southern hemisphere.

Another ambitious project in photometric research was taking shape at the same time. Led by the Paris Observatory, the *Carte du Ciel* made observations in several spots in the southern hemisphere—South Africa, Australia, and in the South American cities of Rio de Janeiro, Rio de La Plata, and Santiago.[23] But the project was later abandoned without having accomplished its goals. It has been said that it was "one of the largest

hemisphere." D.O. Mills Expedition—Arrangements, December 9, 1900. UA 36. Ser 04. Box 8 Folder 1, *Lick Observatory Records*, Special Collections, Mary Lea Shane Archives, University of California Santa Cruz.

[19] Campbell, "Organization and History," 5.

[20] Beverly S. Hard, *Notas sobre los directores de la expedición de D. O. Mills a Chile*, w/d, code 08R-0101, Box 416. Archivo Histórico Universidad Católica de Chile.

[21] Hilmar W. Duerbeck, "National and International Astronomical Activities in Chile," *Astronomical Society of the Pacific Conference Series*, no. 292 (2003): 9.

[22] Heber D. Curtis, "Astronomical Problems of the Southern Hemisphere," *Publications of the Astronomical Society of the Pacific* 21, no. 129 (December 1909): 232.

[23] Albert Gustavus Winterhalter, *The International Astrophotographic Congress and a Visit to Certain European Observatories and Other Institutions* (Washington: Government Printing Office, 1889), 68.

and most ambitious programs ever attempted in international science and by most accounts an embarrassing failure."[24] Unable to foresee the results of *Carte du Ciel*, at the time many observers in the United States argued that the absence of American institutions in the endeavor was a big mistake. More skeptical astronomers pressed for a separate southern hemisphere project in which North American institutions would be completely in charge. Such was the scenario in which Campbell's aspirations could come true.

An observatory in the south would allow astronomers to study the whole sky. Curtis, who would be in charge of the Mills Expedition in Chile a few years later, explained:

> Thirty degrees to the south of the celestial equator to the south pole we have a region amounting to about one fourth of the entire sky which, relatively to the northern skies, was almost as much a terra incognita seventy-five years ago as was Central Africa at the same date, and which today contains many virgin fields which offer rich returns to the exploring astronomer.[25]

Those virgin skies motivated the California astronomers to go far south. But they needed to determine the precise location. As Campbell wrote, "The question of a satisfactory location for the observing station was a most important one. Peruvian stations are too far north for the program of work undertaken; and, except near the sea level, the diurnal range of temperature is large."[26] The distance in navigation and communication with South Africa or Australia, despite being British colonies, could also be a problem. The choice of Santiago de Chile to install the observatory was probably cheaper than in Africa, because the astronomers travelled by sea almost directly from California to Valparaíso, and from there they could reach Santiago by train.[27] Other factors of symbolic nature that may have influenced the decision were James Lick's previous trips to South America, as well as the already close relationship

[24] John B. Hearnshaw, "Photometry, Astronomical," *History of Astronomy: An Encyclopedia*, ed. John Lankford (New York: Garland Publishing, 1997), 339.

[25] Curtis, "Astronomical Problems," 232.

[26] Campbell, "Organization and History," 7.

[27] Ricardo Leiva, "Atacama: a 100 años del informe Curtis," *Revista de Humanidades* 3 (2010): 15.

with Chilean politicians. Indeed, a few years earlier, a small, transitory mission from the Lick Observatory had gone to Chile to observe the April 1893 eclipse.[28]

Whatever their reasons may have been, by 1902 the decision was made. The California press reported on the adventure, the science, and the long miles to travel to Chile. The *San Francisco Call* announced that "Astronomers' Expedition to Leave Lick Observatory Soon for Chile" and detailed the tasks, equipment, and preparations.[29] Having decided on a country, astronomers now had to have an idea where to go in Chile. Santiago, the capital city, seemed to be the best option to Campbell: "The vicinity of Santiago de Chile seemed to be more promising than any other location, taking into account the massive character of the apparatus, the requirements of construction, and the fact that the expedition would probably terminate in less than three years."[30]

Besides, climatic reasons reinforced their decision for, as he observed, central Chile had a fair proportion of clear skies, but one concern was the temperature in Santiago underwent a large diurnal change. Changes in temperature between day and night limited the spectrograph's operation. Fortunately, though, a "much lower range of temperature would be found on the summits of the hills which spring up to an altitude of a few hundred feet from the floor of the valley."[31] Contrary to what one might think, it was not the atmosphere's humidity or urban light conditions that made them look to Santiago's hills, but rather the daily thermal fluctuation. Furthermore, for the moment, it was important to settle in or near a city. The few urban spaces in Chile were found along the Santiago—Valparaíso axis, and another in the south—Concepción, discarded because of its cloudy skies. All in all, they concluded, the advantages of a city near at hand, as a source of material and labor supply during the construction period, all favored Santiago. Once the place was decided, and the funding was secured, the Southern Station of the Lick Observatory became real.

[28] "Lick Observatory Eclipse Station in Chile," *Publications of the Astronomical Society of the Pacific* 6, no. 38 (1894): 290–294.

[29] *The San Francisco Call*, August 27, 1902, 14.

[30] Campbell, "Organization and History," 8.

[31] Campbell, "Organization and History," 8.

2.3 Going South: American Scientists and Chilean Authorities

As in so many other events, individual challenges had a profound and decisive impact on the trajectory of this history. Between 1894 and 1900, when an expedition to the southern hemisphere was no more than a yearning, Campbell hoped to lead it in person. The trip was a scientific mission, but one born from strong personal desire. However, by 1900, Campbell's position in the Lick Observatory had changed and his direct participation in the southern expedition was in doubt. The Regents of the University at that time assigned him new duties, and appointed him as the Director of the Observatory. A change of plan was necessary, and fortunately, Mr. Wright who had assisted him in the work with the Mills Spectrograph from 1897, was both able and willing to take charge of the expedition.[32] Despite the changes in his role at the observatory, Campbell still believed he could supervise, if not all of the research, at least the telescope's installation.

Traveling thousands of miles with complex instruments that needed to be moved and assembled in Chile was a major responsibility. Therefore, he expected to accompany the expedition, in order to select the site, to install the apparatus, and to see the first results. But, alas, he could not. His journey "was prevented almost at the last moment by a severe physical injury received while preparing for the optician's final work on the hyperbolic mirror, and the tests of the two mirrors in combination."[33] The accident was deeply personal for, as he wrote, "the expedition to Chile has been one of my hobbies for the past eight years, and it is unnecessary to say that my disappointment in not being able to go with the expedition is of the most intense kind. Both the San Diego and the San Jose doctors discourage me from having any idea that I can go."[34] Although Campbell remained in charge of the Lick Observatory in California, he was not going to live out his long dream. The "southern twin" would be in the hands of Wright, Campbell's former assistant, "who has ably assisted in the observations with Mills Spectrograph at the

[32] Campbell, "Organization and History," 7.

[33] Campbell, "Organization and History," 7.

[34] D.O. Mills Expedition, Correspondence—W.W. Campbell to Robert Bruce, February 13, 1903. UA 36. Ser 04. Box 8 Folder 5, *Lick Observatory Records*, Special Collections, Mary Lea Shane Archives, University of California Santa Cruz.

Lick Observatory in the last six years, and to whose valuable suggestions and help much of the success of the work is due."[35]

Campbell also made clear that the Chilean government was ready to further the expedition's purposes in every way possible. The host country's participation was, of course, crucial. Before the trip, in 1902, the Lick astronomers had secured Chile's support. The US representative in Chile, Henry Wilson, directly wrote to Eliodoro Yañez, Minister of Foreign Relations, confirming the assistance:

> The Government of Chile extends certain facilities and favors to an astronomical expedition which is to come to Chile in January next, and will probably locate near Santiago, for the observance of that part of the sky inaccessible from Mount Hamilton, California. It is needless to say how pleased my government would be if the Government of Chile would aid the expedition and Professor Campbell in attaining their object, and I have the honor to request your excellency to inform me if the privileges mentioned in Professor Campbell's letter herewith will be granted.[36]

Minister Yañez replied to the American request with enthusiasm, offering both administrative and economic support. He confirmed the terms of Campbell's letter and added that he had sent instructions to the Ministry of Public Instruction. Furthermore, the Ministry reiterated the Chilean government's satisfaction of hosting the American expedition and confirmed the commitment to offer "all the facilities in its power." At the same time, Yañez informed, that Chilean laws exonerate establishments dedicated to studying and teaching—and presumably to research as well- from all fiscal and municipal taxes. He also stated there would be no problem in furnishing the expedition with maps and data, and assistance to settling in the place they chose.[37] In this way, governments, politicians, and authorities became involved themselves in the astronomers' adventure in faraway Chile. Now, preparations could begin.

[35] Campbell, "A Brief Account," 75.

[36] D.O. Mills Expedition—Arrangements, November 8, 1902. H. Wilson to E. Yañez. UA 36. Ser 04, Box 8 Folder 1, *Lick Observatory Records*, Special Collections, Mary Lea Shane Archives, University of California Santa Cruz.

[37] D.O. Mills Expedition—Arrangements, Yañez to Wilson, November 13, 1902. UA 36. Ser 04. Box 8 Folder 1, *Lick Observatory Records*, Special Collections, Mary Lea Shane Archives, University of California Santa Cruz.

2.4 Building and Shipping the Equipment

The conditions and resources in Santiago favored the decision to settle there. It was important to ensure that astronomers could find materials they needed to build the observatory, as well as supplies to operate the spectrograph. In addition, the team would need qualified labor to aid in these tasks and workers for the construction. An American acquaintance who resided in Chile, W. R. Hudson, explained to Campbell what he could expect to find.

> In Santiago, brick, cement, timber, and lumber of all kinds can be easily and comparatively cheaply obtained. Good carpenters are also to be found, and I think that you would be able to get any work done that you required, promptly and efficiently. There is no necessity whatever for having the United States Consul contracting for the timber in advance; 6 × 6 inches lumber is quite a common size in the Valparaíso wood yards. You can get ordinary machinery repaired, and even manufactured, in Valparaíso. You refer to your apparatus, but I do not know how much you make the word extend. I do not think, for instance, that you could get lenses of any size to replace breakages, but you could get ordinary brass work done, and there are good chronometer makers who would no doubt be able to assist.[38]

Mr. Hudson advised Campbell to inform the Chilean authorities about the expedition, so that they could prepare for the astronomers' arrival. He suggested to contact the United States Consul and ask him to advise the Chilean authorities that the University of California has this expedition in view. In addition to the material resources, the astronomers had to think of climate, geography, and the city's technological conditions. Hudson noted that,

> The nights in Santiago are very clear and there is almost no fog, except in the months beginning May, June and July, when the rainy days are almost invariably accompanied by fog. There are water works in the City, from

[38]D.O. Mills Expedition—Arrangements, January 14, 1901. W.R. Hudson to W.W. Campbell. UA 36. Ser 04. Box 8 Folder 1, *Lick Observatory Records*, Special Collections, Mary Lea Shane Archives, University of California Santa Cruz.

which no doubt water could be secured, and the electric light is used, but to a limited extent only. I am not in a position to say whether you will be able to secure a sufficient supply for motor power.[39]

To know about these characteristics was essential for the project's success. Expedition leaders needed to go through all the details since there were no similar observatories in the southern hemisphere.

Most of the observatory's pieces were ordered and carefully built by companies in the United States. The original mirror came from Pennsylvania but broke, thus imperiling the expedition's future. Another one was ordered from Allegheny, where there was an observatory and qualified men to work on those instruments. Success in building the pieces was described by the *New York Times*: "Mirrors for the stars. The reflecting telescope mirrors for use with the D.O. Mills Expedition to Chile, recently completed by an Allegheny, Penn. firm have reached Mount Hamilton in perfect conditions."[40] Many companies and individuals were involved in the construction of the telescope destined for Chile. The telescope's mount was designed in San Francisco and built in Los Angeles; the optical pieces arrived from Pennsylvania; and the dome was built and shipped from Cleveland.[41] Extreme care was vital. Once in South America, the astronomers could not count on the availability of materials to install and operate the telescope.

There was also a relevant cultural factor to be considered. The cosmopolitan nature of Valparaíso meant that there were Americans both in the port and Santiago. In fact, the astronomers hoped that presence of the Williamson, Balfour Co. could be helpful for the expedition. Obviously, the astronomers would have to make assertive decisions regarding a place in Santiago. "If I might make a suggestion," Hudson wrote, "I would say that it would be well to secure ground, or permission to

[39] D.O. Mills Expedition—Arrangements, January 14, 1901. W.R. Hudson to W.W. Campbell. UA 36. Ser 04. Box 8 Folder 1, *Lick Observatory Records*, Special Collections, Mary Lea Shane Archives, University of California Santa Cruz.

[40] *New York Times*, July 13, 1902.

[41] John A. Brasher Co., St. Gobain Plaste Glass Co., both from Pennsylvania, participated, as well as the Allegheny Observatory, Harron, Richard & McCone Co. from San Francisco, Fulton Engine Works of Los Angeles, Warner ¢ Swasey de Cleveland among others. Campbell, "Organization and History," 5ff.

use ground, suitable for putting up the building, which you evidently require."[42] Another urgent task was to secure an electric supply. In a September 1902 letter, the expedition planners discussed the subject. Some agents for the West Coast of South America informed that the majority of the electric plants in Chile were the 220/440 volt three-wire system.[43]

Once the decisions regarding the southern station location were made, the goal became even more ambitious. They had already planned the starting point: the telescope and the dome would be mounted on the summit of one of the low hills near Santiago, and the mission would work there for two or three years. If at first they had planned for measuring about 170 radial velocities, soon they increased the number of measurements and settled the joint work with the telescopes in California.

> It is planned to measure the velocities of three of four hundred of the brighter stars in the southern sky. The photographs will be sent to the Lick Observatory for measurement and discussion. It is confidently expected that a combination of the spectroscopic results obtained at Mt. Hamilton and Santiago will lead to a very satisfactory solution of the troublesome problem of the motion of our solar system through space.[44]

The expedition to Chile was real. The goal expanded and became more ambitious, and the astronomers were getting ready to go to the southern hemisphere, taking care of all the details of this delicate and innovative expedition. Chile was an exotic distant land. The country inspired the astronomers' confidence. But the prospect of traveling thousands of miles still made them dizzy, despite some previous short expeditions to observe eclipses. They were about to start a new life and a new scientific endeavor. The next step was to travel and settle in Chile.

[42] D.O. Mills Expedition—Arrangements, January 14, 1901. W.R. Hudson to W.W. Campbell. UA 36. Ser 04. Box 8 Folder 1, *Lick Observatory Records*, Special Collections, Mary Lea Shane Archives, University of California Santa Cruz.

[43] D.O. Mills Expedition—Arrangements, September 22, 1902. M.A. Oudin to Campbell. UA 36. Ser 04. Box 8 Folder 1, *Lick Observatory Records*, Special Collections, Mary Lea Shane Archives, University of California Santa Cruz.

[44] Campbell, "A Brief Account," 74.

Bibliography

Campbell, William W. "A Brief Account of the D.O. Mills Expedition to Chile," *Publications of the Astronomical Society of the Pacific* 15, no. 89 (April 1903): 70–75.

Campbell, William W. "Organization and History of the D.O. Mills Expedition to the Southern Hemisphere," *D.O. Mills Expedition, Cerro San Cristóbal, Santiago, Chile, from Its Organization in December 1900 to March 1906*, edited by W.W. Campbell and W.H. Wright, Vol. IX, 5–14. Sacramento: W.W. Shannon Superintendent State Printing, 1907.

Campbell, William W. "Notes from Pacific Coast Observatories," *Publications of the Astronomical Society of the Pacific* 40, no. 236 (1928): 249–252.

Curtis, Heber D. "Astronomical Problems of the Southern Hemisphere," *Publications of the Astronomical Society of the Pacific* 21, no. 129 (December 1909): 231–244.

"Darius Ogden Mills," *New York Times*, November 27, 1898.

"D.O. Mill," *National Bank of D.O. Mills and Company Records, 1847–1927*, Manuscript 368–378, California State Library.

D.O. Mills Expedition, UA 36. Ser 04. Boxes 1 to 12, *Lick Observatory Records*, Special Collections, Mary Lea Shane Archives, University of California Santa Cruz.

Duerbeck, Hilmar W. "National and International Astronomical Activities in Chile," *Astronomical Society of the Pacific Conference Series* 292 (2003): 3–20.

Hard, Beverly. *Notas sobre los directores de la expedición de D. O. Mills a Chile*, w/d, code 08R-0101, Box 416, Archivo Histórico Universidad Católica de Chile.

Hearnshaw, John B. "Photometry, Astronomical," *History of Astronomy: An Encyclopedia*, ed. John Lankford. New York: Garland Publishing, 1997.

Karl, Barry D., and Stanley N. Katz. "The American Private Philanthropic Foundation and the Public Sphere 1890–1930," *Minerva* 19, no. 2 (June 1981): 236–270.

Leiva, Ricardo. "Atacama: a 100 años del informe Curtis," *Revista de Humanidades* 3 (2010): 10–22.

"Lick Observatory Eclipse Station in Chile," *Publications of the Astronomical Society of the Pacific* 6, no. 38 (1894): 290–294.

New York Times, July, 13, 1902.

Ridpath, Ian. *Dictionary of Astronomy*. Oxford: Oxford University Press, 1997.

Sterken, Christiaan. "Radial Velocity Observations with the 36" Telescope at Cerro San Cristóbal, Santiago," *The ESO Messenger* 28 (1982): 12–13.

Swett Marden, Orison. "Darius Ogden Mills: A Self-Made Man Who Strives to Give Others a Chance," *Little Visits with Great Americans: Or Success, Ideals, and How to Attain Them*, Vol. 1. New York: The Success Company, 1905.

The San Francisco Call, August 27, 1902.

Winterhalter, Albert Gustavus. *The International Astrophotographic Congress and a Visit to Certain European Observatories and Other Institutions.* Washington: Government Printing Office, 1889.

CHAPTER 3

The Astronomers' Arrival and Challenges of Getting Settled in Chile

Abstract This chapter presents Americans' arrival to Chile and their first impressions: the country was supposed to be a quiet, trustworthy one, but a strike jeopardized the entire mission. This event shows an interesting cultural gap; through the eyes of the scientists we can see the 1900s Santiago, while astronomers chose the place to build the observatory, and work began. This chapter expands on the idea of this story as one of mirrors, reflecting habits, lifestyles and mentalities of two different societies. These societies behaved very differently, but oddly had similar landscapes, and even a similar seismic pattern. While the Americans were getting used to the southern country, they were also making impressive progress in astronomical science, mainly to solve the so-called sidereal problem.

Keywords Chile in 1900s · San Cristóbal Hill · 1906 earthquakes

For seven years, William Campbell had been nurturing the idea of conducting astronomical observations in the southern hemisphere. Now that the project was becoming concrete, it had to be carefully prepared. The planning was about money, but also much more. Among many other activities, Campbell and his team had to contact local authorities,

B. K. Silva, *Astronomy at the Turn of the Twentieth Century in Chile and the United States*, Palgrave Studies in the History of Science and Technology, https://doi.org/10.1007/978-3-030-17712-6_3

find an appropriate spot in the central area of the country, and design and build the instruments. Following negotiations over budgets with different shipping companies, the astronomers settled on the steamer *Perú*, which set sail from San Francisco on February 28, 1903. The scientific team took the name of the benefactor who made the adventure possible: The D.O. Mills Astronomical Expedition from the Lick Observatory, University of California. The purpose of the expedition, as they explained was "to measure spectroscopically the line-of-sight velocities of the naked-eye stars in the Southern Hemisphere which are not visible at Mt. Hamilton."[1]

The astronomers' departure inaugurated the expedition. They would spend several weeks on the deep blue ocean that was more mysterious than the Atlantic. The trip presaged their experiences in Chile. It was the beginning of the unknown, which, at the same time, provoked a strange familiarity. The same cold, strong ocean with its tough waves and currents would take them from the California coastline to distant shores on the other side of the planet. They would increase their knowledge of the celestial bodies that rose above the lights that reflected in the water, in the dark immensity of the nights aboard the steamship Peru. The calm and patience of the scientific spirit was also present during the trip. As Campbell learned from the acting astronomer, Wright, "We met with few adventures during the passage which are not to be accounted normal to travel in those parts of the world, and none that call for record in these pages."[2] Thus began a vast correspondence between north and south that would last for several years. When Campbell read this first letter, he may have felt somewhat disappointed that he was not aboard the steamship, leading the adventure. After the ravages of the waves, the tides, the seasickness, and the long weeks at sea, the arrival in Chile would be much more unexpected than they had anticipated.

[1]American Association of the Advancement of Science, "The D.O. Mills Astronomical Expedition," *Science* 17, New Series, no. 428 (March 1903): 436.

[2]William H. Wright, "Introductory Account of the D.O. Mills Expedition," *D.O. Mills Expedition Cerro San Cristóbal, Santiago, Chile from Its Organization in December 1900 to March 1906*, ed. W.W. Campbell and W.H. Wright, Vol. IX (Sacramento: W. W. Shannon Superintendent State Printing, 1907), 15.

3.1 Arriving to Valparaíso in 1903

Wright and his family had probably heard that Chile—compared at least to other, supposedly chaotic, convulsive, and uncivilized Latin American countries—was reliable. And stability was the first impression that greeted Wright in Chile. On April 18, the steamer reached the Chilean shore and Wright immediately wrote to his boss in California. "We found on landing that this country is enjoying some, if not all, of the blessings of an advanced civilization."[3] Nevertheless, surprise soon took over this initial impression. "All the lighter hands and cargo hands 'walked out' on a 'straique' during Thursday and Friday, and the port is tied up tighter than was San Francisco under similar circumstances. We landed the mirror on Monday by means of a row boat."[4]

It was the era of the so-called social question. The world phenomenon was a reaction to the profound changes in the living conditions of the working classes as peasants migrated to cities looking for better lives.[5] The social question referred to the social and ideological changes that sprang from industrialization and urbanization.[6] It was about new wage-based labor relationships, as well as precarious housing and health conditions for workers. Social actors created new organizations to defend workers against deplorable working and living conditions. Stunned oligarchs the world-over responded with repression.[7]

Wright was describing the 1903 Port Strike, one of the three major strikes of the early twentieth century in Chile. It was no surprise that it was taking place in the "Jewel of the Pacific," as Valparaíso was known. A mandatory stop for interoceanic navigation, the port was showing the strains of its modernizing aspirations. The constant stream of travelers and migrants stimulated its rhythm.[8] In April 1903, groups of stevedores

[3] Beverly Hard, "Notes on the Directors of the D.O. Mills Expedition to Chile," 1972. Expeditions, UA 36. Ser 04. Box 12 Folder 6, *Lick Observatory Records*, Special Collections, Mary Lea Shane Archives, University of California Santa Cruz, 3.

[4] Hard, "Notes on the Directors," 3.

[5] Bárbara Silva, *Identidad y nación. Patria Vieja, Centenario y Bicentenario* (Santiago: Lom, 2008), 109ff.

[6] James Morris, *Las elites, los intelectuales y el consenso* (Santiago: Pacifico, 1967), 79.

[7] Mario Garcés, *Crisis social y motines populares en el 1900* (Santiago: Documentas, 1991), 234.

[8] Bárbara Silva and Josefina Cabrera, *Chile, 100 días en la historia del siglo XX* (Santiago: Planeta, 2015), 58.

paralyzed their work to demand better working conditions. In May, massive protests shook the port city. Looting and torching of buildings, including the offices of the newspaper *El Mercurio* and several credit houses, ensued. "Unfortunately, the day of our arrival in Valparaíso," Wright noted, "marked the outbreak of an extended strike of launch hands, that culminated a month later in a most serious riot, in which a number of business buildings were burned, and over a hundred men are said to be killed in the streets."[9]

The strike hindered the unloading of the delicate equipment, a severe concern for Wright, the head of the expedition. "During this interval of a month and over, practically no merchandise could be unloaded from lighters, nor moved to the Custom House, nor along the waterfront. Most of our belongings were unloaded into lighters and left to bob [sic] around on the bay."[10] The unexpected strike could have been fatal for the expedition, the success of which depended, almost entirely, on the equipment they had brought with them. After investing so much time and energy in the expedition's preparation, a strike at the port was the last scenario they could have imagined—much less in a place like Chile, which different foreigners in the country had assured them was calm and reliable. The mirrors, the telescope, and the dome had endured the transoceanic journey. Now, in Valparaíso, all their efforts (not to mention the thousands of dollars invested) could be lost amid absolute chaos.

But luck was on Mr. Wright's side. "I was fortunate enough to be able to get the mirrors and our telescope, weighing with their packing some nine hundred pounds, into, or rather across the gunwales of a passenger rowboat, and thus balanced we rowed them gently to the shore."[11] Suddenly, Wright became an improvised dockworker, his astronomical abilities set aside in favor of his rowing skills. Just a few months earlier, on Mount Hamilton's summit, all this history would have seemed unlikely. But there he was, the gringo: half scientist, half boatman, trying to protect the future of the southern observatory with the strength of his arms.

[9] Wright, "Introductory Account," 16.

[10] Wright, "Introductory Account," 15–16.

[11] Wright, "Introductory Account," 15–16.

They managed to rescue the most sensitive equipment, but other pieces were still adrift in Valparaíso Bay. From a sit-down strike, the movement ascended to violence. The police repressed the protest with even greater violence. Then, the conflict was suspended for a short time that was useful for the astronomers. Wright informed Campbell that the strike ended in late May, and fortunately, their equipment was safe. As soon as they could, they transferred it to Santiago and got the work of construction under way.[12]

3.2 Looking for the Place

Once they had negotiated the difficulties of the strike, the expedition set course to Santiago. Wright knew the distance to be approximately 120 miles by train. It seemed appropriate to look for whatever possible comforts the capital could offer. On the way, Wright's assistant, Dr. Palmer, made a detour to the town Los Andes, in the Aconcagua Valley, to see if there were any advantages in that region for astronomical observation, but there were not.[13] Later, they spent one month looking for the best possible location for the observatory in the central valley, in the Santiago area.

The secretary of the United States Legation in Chile informed Dr. Campbell of the expedition's arrival. "I began to think the expedition was doomed not to arrive, when it suddenly appeared, like an unknown star, and I am very pleased to say that Mr. Wright and his family have found a house to live in, where I think they will be comfortable, until they locate a place for the observatory."[14] Once settled, Wright began to evaluate the city's suitability for the observatory. Becoming acquainted with Santiago was his main task and his description of the sites he visited was accordingly exhaustive.

[12]Wright, "Introductory Account," 21.

[13]Wright, "Introductory Account," 16.

[14]D.O. Mills Expedition—Arrangements 1901–1903, May 1, 1903, UA 36. Ser 04. Box 8 Folder 1. *Lick Observatory Records*, Special Collections, Mary Lea Shane Archives, University of California Santa Cruz.

> The city is situated five or six miles from the precipitous slopes of the Andes, or Cordilleras, as they are more commonly called, on a mountain-locked plain sloping gently toward the west. To the northeast, and towering over the valley, is a fine group of mountains known as Las Cordilleras de Los Condes, the culmination peak of which, Cerro del Plomo, is a magnificent mass of rock over nineteen thousand feet high, and perpetually covered with snow. On all sides are mountain ranges, which are secondary importance only because they are associated with giants. Within the city, a number of rocky hillocks rise abruptly from the alluvial deposit upon which the city is built, while to the northeast a long precipitous spur reaches out from the Cordilleras and after dipping to the level of the valley rises again and culminates in a hill, somewhat over a thousand feet high, called Cerro San Cristóbal. The country surrounding Santiago is made fertile by irrigation and is noted for its productiveness.[15]

These observations of Santiago were fundamental. The primary duty was to choose the best place for the observatory. Wright was to live and work there for the next two or three years. They discussed settling in the southern part of the city as US diplomats had recommended. They suggested San Bernardo, at a station on the southern railway line, where there was a hill known as Cerro Negro. An English gentleman, Carlos Rogers, was trying to buy it, and seemed "anxious to assist the expedition, and to put this hill at their service."[16] The secretary of the United States Legation in Chile seemed to know in detail the needs of the expedition and what kind of place they were looking for. Referring to Cerro Negro, the description contained aspects of the domestic and labor needs of the scientists.

> There is a house with four rooms on top of the hill, which could be used for officers or workrooms, and also a house at the base of the hill, where Mrs. Wright could reside with her husband and sister. There is also electricity at San Bernardo, which is needed in the observatory, for keeping the instruments in an even temperature. If Don Carlos can get control of

[15] Wright, "Introductory Account," 16.

[16] D.O. Mills Expedition—Arrangements 1901–1903, May 1, 1903, UA 36. Ser 04. Box 8 Folder 1. *Lick Observatory Records*, Special Collections, Mary Lea Shane Archives, University of California Santa Cruz.

> this hill, I felt certain it will answer the purposes of the expedition; but, of course, you will be more fully advised by your lieutenant, Mr. Wright, whom we like very much, and his wife, also.[17]

In addition to the evidence of how the initial interpersonal relations began to come together, the search for the observatory's location showed other dimensions. These explorations made clear that foreigners were already integrated into Chilean society and that they could act as a social bridge for the scientists.

Although there are no conclusive records that explain why that hill in San Bernardo was not chosen, the distance within the city could have been a factor. The search also faced some climatic obstacles, since it was the beginning of the wet season. It was of course impossible to judge observing conditions if it was cloudy and rainy.[18] Finally, after avoiding the rain and studying possible places, they chose to settle in the northern part of the capital, in Cerro San Cristóbal. Finding a place of a certain altitude was expected; the original Lick Observatory was located on Mount Hamilton near San Jose and several other scientific investigations were carried out on hills near urban centers of different sizes.[19] Features of San Cristóbal were very precise:

> A month was devoted to the survey and study of the most promising locations. The summit of *Cerro San Cristóbal* was selected. This peak is located at the northeastern edge of Santiago, about 2.3 kilometers from the central plaza of the city. The Observatory site is 920 feet (280 meters) above the *Plaza de Armas*, 885 feet (270 meters) above the foot of the *cerro*, and 2755 feet (840 meters) above sea level.[20]

[17] D.O. Mills Expedition—Arrangements 1901–1903, May 1, 1903, UA 36. Ser 04. Box 8 Folder 1. *Lick Observatory Records*, Special Collections, Mary Lea Shane Archives, University of California Santa Cruz.

[18] Wright, "Introductory Account," 19.

[19] Charlotte Bigg, David Aubin, and Philipp Felsch, "Introduction: The Laboratory of Nature—Science in the Mountains," *Science in Context* 22, no. 3 (2009): 311–321.

[20] William W. Campbell, "Organization and History of the D.O. Mills Expedition to the Southern Hemisphere," *D.O. Mills Expedition Cerro San Cristóbal, Santiago, Chile from Its Organization in December 1900 to March 1906*, ed. W.W. Campbell and W.H. Wright, Vol. IX (Sacramento: W. W. Shannon Superintendent State Printing, 1907), 8.

Cerro San Cristóbal seemed appropriate for several reasons. It was in the city's suburbs and therefore offered a balance of tranquility and connectedness. San Cristóbal was outside the city's boundaries, and in the city there was little artificial light, besides Santiago was close to the Port of Valparaíso and the railroad allowed transporting the equipment easily.[21] This was critical. Although the astronomers would settle in Chile, they would need to be in constant connection with California in order to maintain supplies, replacement parts, and communication with the Lick Observatory. Furthermore, they needed basic visibility conditions and San Cristóbal seemed to offer them. "Its summit is nearly always well above the dust and haze of the city, and it is free from the ground fog which frequently covers a large part of the valley before dawn on summer mornings."[22] Additionally, they had to worry about the temperature range and its difference between day and night, for that could seriously affect the spectrograph's operation, and the diurnal range in temperature was much less on the hill than on the plain. As much as they needed to be above the fog and the city lights, the location also had to be accessible. It had to be low enough to climb easily, but safe from the blasting in the nearby rock quarry and from a shooting range.[23] Clearly the hill, although near the city, retained the rural air of a rock quarry where hunters went to shoot.

Wright was in charge of the delicate task of selecting a place, and he had to make all these fundamental arrangements "without the present help of his injured leader, although in spite of his injury and his disappointment Dr. Campbell was able to advise Dr. Wright almost as if he had been present."[24] Certainly, Wright's tasks were not limited to installing the equipment; he also was to prepare the land, assemble the pieces, and secure the electric supply. They broke ground for the Observatory buildings on May 27, 1903, and the installation was completed on September 11, with the exception of the electric line, which was completed in October.[25] It was a point of pride to have been able

[21] Ricardo Leiva, "Atacama: a 100 años del informe Curtis," *Revista de Humanidades* 3 (2010): 15.

[22] Wright, "Introductory Account," 19.

[23] Hard, "Notes on the Directors," 2.

[24] Hard, "Notes on the Directors," 2.

[25] Wright, "Introductory Account," 21.

to negotiate all the obstacles that the construction of an observatory implied. "Notwithstanding the difficulty of installing an observatory on the mountains in the winter season, the dome, telescope, office building, electric power-line to the valley, and the many accessories were all completed early in October (1903)."[26] It was sophisticated, delicate equipment, but Campbell's beliefs were conclusive when deciding for it, even if it meant traveling with a spectrograph to the faraway south:

> The low cost of the Chile 36-inch reflecting telescope (5,500 in California) is in striking contrast with the 18-fold greater cost of the 36-inch refractor, though I do not doubt that the same or equally good locations, and with good silver coats on the mirrors, the two instruments would be almost equally efficient in radial-velocities determinations. Necessarily this was the most important factor in making my decision to employ a reflecting telescope.[27]

Work on the southern station of the Lick Observatory thus began. A few months later, the astronomical research started. On September 11, 1903, 280 meters above the Plaza de Armas of Santiago, the 92-centimeter mirror saw light for the first time. The observatory remained in operation under Lick management until 1928, taking thousands of spectrograms of radial velocities with the Mills Spectrograph.[28] What had begun as a two or three-year expedition soon became a much more ambitious enterprise.

3.3 First Relationships with Chileans: Science or Politics?

Besides dealing with all the unexpected difficulties that came with the construction work and task of assembling the pieces, the American mission had to try to be involved socially with Chilean society. Their stay was not supposed to be long, but they still needed to bond with local

[26] Campbell, "Organization and History," 8.

[27] Campbell, "Organization and History," 11.

[28] Leiva, "Atacama: a 100 años," 15.

residents and even more with the authorities who could ease their tasks; fortunately, Chileans welcomed the expedition. Politically organized into a "parliamentarian system," Chilean society longed to be closer to the developed societies of the North Atlantic world.[29]

Wright wrote that upon their arrival in Santiago, they were presented to His Excellency Henry M. Wilson—the American Minister—, the Secretary of the Legation Norman Hutchinson, and Señor Don Manuel Foster, Assistant Secretary of Foreign Affairs.[30] Manuel Foster already appeared in the history of the observatory. The assistant secretary could not have suspected how decisive his presence would be for the future of the scientific initiative, nor that years later the observatory would be renamed after him.[31] Welcoming the expedition can be attributed to his own initiative, and from the very beginning expressed the desire of the government to aid the mission by all means within his power.[32] Moreover, Foster seemed to have been involved even in a personal level, as Mr. Hutchinson informed Campbell:

> I personally met Mr. Wright at the station, on his arrival in Santiago, and saw him to the hotel in the Minister's carriage, which he put at Mr. Wright's disposal. Both the Minister and myself have had Mr. Wright and his family to dine, and on both occasions, we had present the Sub-Secretary of Foreign Relations, Senor Don Manuel Foster Recabarren, who took an interest in the expedition, as he has since first advised by this Legation of its expected visit to Chile. Mr. Foster has offered every facility, and railway passes, and even the use of his country house and grounds gratis.[33]

At first, Chilean help came in concrete actions that facilitated the expedition's arrival. Wright addressed the generous disposition of the Supreme Government of Chile, as well as "a number of courtesies"

[29] Silva, *Identidad y nación*, 79–85.

[30] Wright, "Introductory Account," 16.

[31] Twenty-five years later, when the Americans were about to leave Chile, Manuel Foster bought the observatory and donated it to the Catholic University. The University named the observatory after him.

[32] Wright, "Introductory Account," 16.

[33] D.O. Mills Expedition—Arrangements 1901–1903, May 1, 1903, UA 36. Ser 04. Box 8 Folder 1. *Lick Observatory Records*, Special Collections, Mary Lea Shane Archives, University of California Santa Cruz.

demonstrated by Chileans in general. Moreover, Wright gave details about this "generous disposition": they did not have to pay for any custom fees for all their freight and luggage; they also got passes on the State railroads and letters of introduction signed by the Minister of Public Works, and the government placed two cars—quite a luxury back then—for transporting their equipment from Valparaíso to Santiago.[34]

The Chilean government's favorable attitude concerning the expedition pointed toward the support of Manuel Foster.[35] American astronomers thus began to touch base, legally and personally, with members of the Chilean government and to learn about the functioning of the young republic at the end of the world.

Local officials not only facilitated the installation procedures for the expedition; they also promptly understood the need to protect and take care of these foreigners, lest their stay became a diplomatic incident—the port strike had already been enough. When the team was still studying sites for the observatory, Wright took careful note of the Santiago policing provisions. "The Prefect of Police of the Municipality of Santiago offered us the services of a policeman for protections, it having been brought to his attention that the visiting astronomers were in the habit of taking thermometer readings on isolated hills and peaks during the lonely watches of the night."[36] Even if the police did not understand the astronomer's work, the Americans would soon understand how significant this offer was. Considering their valuable—indeed, irreplaceable—equipment they had installed on the summit of San Cristóbal Hill, the fact that "every afternoon a policeman leaves the *cuartel* in Santiago and climbs the nine hundred feet to the Observatory," proved indispensable.[37]

William H. Wright and his assistant Mr. Palmer settled down in Santiago in 1903. Little by little, they and their families integrated into the remote society of Santiago. They began inhabiting the desolate San Cristóbal—avoiding the quarry, and the hunters, and the city lights that,

[34] Wright, "Introductory Account," 18.

[35] D.O. Mills Expedition—Arrangements 1901–1903, May 1, 1903, UA 36. Ser 04. Box 8 Folder 1. *Lick Observatory Records*, Special Collections, Mary Lea Shane Archives, University of California Santa Cruz.

[36] Wright, "Introductory Account," 18.

[37] Wright, "Introductory Account," 18.

though scarce, still radiated a sense of progress sought by the Chileans. Their astronomical tasks of observing radial velocities was always a priority, but they understood the importance of cultivating relationships with social circles and, of course, the small but growing group of local astronomers. The Americans knew about how the Gilliss mission of a half-century before had helped launch Chile's National Astronomical Observatory (NAO).

Although the nature of the two scientific endeavors was very different, the information registered by the NAO could be useful. They said the National Observatory was practically a meteorological observatory, but was indeed very helpful to have weather records as it did, starting in 1861.[38] From the beginning, the Californian astronomers contacted the NAO and tried to relate with local scientists. In fact, they agreed to collaborate directly with NAO Director, the astronomer Albert Obrecht, who extended them a warm welcome and, if needed, a courteous invitation to use the resources of the institution under his charge. The offer was not simple courtesy. In fact, it was very useful to Wright and his team. "We had occasion to accept his invitation many times, and are indebted not only to Director Obrecht, but also to Professor Ernesto Greve, First Astronomer, and to M. Krahnass, the Observatory Meteorologist, for the furnishing of certain much needed meteorological data, and for assistance rendered in many other ways."[39] Obrecht offered the observatory grounds—located in the *Quinta Normal de Agricultura*, a park in the western-lower end of the city—for the American astronomers to use.

The NAO's brick tower could have been useful, for it was just the size required for the dome. Nonetheless, there was a wide diurnal range in temperature, which left little room for hope that a reflecting telescope could be used in that place.[40] Wright's gratitude toward Obrecht was sincere, but his reaction also suggested that, in his view, the NAO did not have a clue what a reflecting telescope was, nor

[38] D.O. Mills Expedition, Correspondence—W.W. Campbell to D.O. Mills. Mount Hamilton, January 21, 1905. UA 36. Ser 04. Box 8 Folder 3, *Lick Observatory Records*, Special Collections, Mary Lea Shane Archives, University of California Santa Cruz.

[39] Wright, "Introductory Account," 18.

[40] Wright, "Introductory Account," 19.

spectroscopy, and even less why the daily temperature range was of crucial relevance. Even so, the hospitable welcoming came not only from politicians and diplomats, but also from Chile's small scientific community.

3.4 The Dome's Opening and the Need to Plan the Future

Shortly after the Americans' arrival to Chile, Campbell began to look into the possibility of extending the expedition's timeframe. The work of building the equipment, moving it to distant lands, and installing every detailed piece of an observatory merited a longer, more fruitful period. He may have harbored such intentions during the preparatory stages—planning, securing two years of financing with Mills, and, finally, seeing the expedition's safe arrival. Still, he demurred. "In forming the general plans and in designing the instrumental equipment," he later wrote, "I had no thought other than the work of the expedition would terminate at the end of two years of actual observation, and the instruments be dismounted and returned to this country."[41] Accordingly, the team designed the instruments and treated them with great care, for their destination was as uncertain as the expedition was costly. Therefore, "the more massive parts of the telescope were kept as light as possible", since the place in Chile might or might not have a road, and they did not know if they would have to climb up and then climb down with the telescope and the dome—that is why at first they covered it with heavy canvas instead of metal.

To the extent that this work would help elucidate the sidereal problem through radial velocities, still a major scientific challenge, it was worth the effort. For Campbell, the measurement of stars' radial velocities was one of the "most difficult problems in modern observational astronomy."[42] In this challenge, astronomers had some antecedents to the Lick Southern Station and therefore understood the many obstacles. Campbell was nonetheless optimistic. "The first serious attempt to measure stellar

[41] Campbell, "Organization and History," 9.

[42] Campbell, "Organization and History," 9.

velocities with a large Cassegrian reflecting telescope, involving many radical departures from conventional forms, would develop its own crop of difficulties to be removed one by one."[43] Largely due their attention to details, the astronomers were going to be able to overcome these challenges. It was therefore reasonable to think of a longer expedition.

In fact, the team's enthusiasm led them to believe it would be necessary to test every piece and every assemblage. This was required to secure the mission's success. For instance, before departing to Chile they hauled all parts of the telescope to Mount Hamilton and assembled them, trying to anticipate and remove any defects and difficulties.[44] Although they tried to foresee any additional obstacle, the future awaiting them was unknown. Despite all tests, the astronomers had to trust that the instruments would function perfectly in faraway Chile. It was a matter of joy to learn that when they assembled the telescope on a summit of a mountain in Chile, thousands of miles away from instrument makers, everything seemed to work out just fine. Indeed, it must have been a great relief to learn that all the preparations had paid off. At least, the observatory worked.

The construction of the observatory on San Cristóbal Hill, during the rainy season, represented another challenge. Nonetheless the accuracy of the observations appeared to match that of the observations made with the Mills spectrograph at Mount Hamilton.[45] With pride and satisfaction, and of course huge relief, Campbell confirmed that the quality of the results obtained in Chile was as reliable as those taken from California. In addition, Campbell underlined how they all had carried out this enormous enterprise with great financial caution. For the Director of Lick the expedition's success merely pointed the way for the future. "The two years of actual observation covered by my estimate expense will be up by next year. Inasmuch as a letter and its answer require nearly three months to go and return, and professor Wright and I must make our plans in advance, it is necessary that the future of the Expedition be determined in the next few weeks."[46]

[43] Campbell, "Organization and History," 9.

[44] Campbell, "Organization and History," 9–10.

[45] Campbell, "Organization and History," 10.

[46] D.O. Mills Expedition, Correspondence—W.W. Campbell to D.O. Mills. Mount Hamilton, January 21, 1905, UA 36. Ser 04. Box 8 Folder 3, *Lick Observatory Records*, Special Collections, Mary Lea Shane Archives, University of California Santa Cruz.

Campbell took great care to inform his benefactor Mills of the success of the southern observations: in fact from 1903 until 1904 they had obtained 330 photographs of star spectra. To that he had to add the variable of climate, which was not helping the radial velocities measurements, as Mr. Wright informed the number of cloudy days that year was greater than in any other year, and the rainfall was three times the average amount. Even with those cloudy nights ratings, he could sustain the financial efficiency of the astronomical expedition, something that was certainly relevant to Mills. "For $24,000 plus 10% or less for unexpected expenses, we have constructed and erected a powerful telescope and spectroscope, a steel dome, small buildings, etc., composing an observatory more efficient than many existing observatories costing several times as much."[47] Campbell thus emphasized his careful management of the budget.

The success of the expedition contrasted with the initial difficulties. As Campbell made that clear, "The unfortunate delays encountered in completing the mirrors for the reflecting telescope made it unavoidable that the Expedition should reach Chile at the beginning of the southern winter's rainy season."[48] In this way, he attributed the expedition's delay to the management in California, not to the tasks undertaken by the Americans upon their arrival in Chile. Yet, he did acknowledge some obstacles, such as the Valparaíso strike and the "storms of winter." In addition, settling in an unfamiliar Latin American country with limited technical capacity brought more challenges that demanded the almost heroic versatility of the astronomers. "The requirements for erecting the Warner & Swansey dome, the reflecting telescope, the electric power line from the valley to the summit, etc., could not be set by skilled labor obtainable in Chile, and the astronomers were called upon to supply this almost entirely themselves."[49] Despite these

[47] D.O. Mills Expedition, Correspondence—W.W. Campbell to D.O. Mills. Mount Hamilton, January 21, 1905, UA 36. Ser 04. Box 8 Folder 3, *Lick Observatory Records*, Special Collections, Mary Lea Shane Archives, University of California Santa Cruz. Emphasis in original.

[48] D.O. Mills Expedition—Arrangements 1901–1903, September 1, 1903. UA 36. Ser 04. Box 8 Folder 1, *Lick Observatory Records*, Special Collections, Mary Lea Shane Archives, University of California Santa Cruz.

[49] D.O. Mills Expedition—Arrangements 1901–1903, September 1, 1903. UA 36. Ser 04. Box 8 Folder 1, *Lick Observatory Records*, Special Collections, Mary Lea Shane Archives, University of California Santa Cruz.

troubles, which Wright and Palmer could not have even imagined, they secured spectrograms for determining the speeds of twenty stars in a couple of months.

The initial outcomes were promising, Wright and Campbell recalled years later. The measurements were trustworthy and the data collected were comprehensive, all of which was crucial to the accounts they offered to Mills and to the validation they expected from their colleagues. In fact, in the three subsequent years, that initial number of plates went up to 800:

> The spectra of all the brighter stars south of declination -25° were photographed, and on the basis of the observations thus secured, a working list of 145 stars was made up. This list includes all the stars whose spectra were found to contain measurable lines. Four spectrograms were obtained of practically every star on the list, and additional ones of many other stars were taken for special purposes. The total number of spectrograms secured by the expedition is 800, of which 676 are of stars on the regular list, 92 are of stars whose spectra were found to be unsuitable for accurate measurement, and the remaining 32 are check plates of Mars, Venus and the Moon.[50]

Despite the vast data they were collecting from unknown southern skies, the number of spectrograms discarded by the expedition was very low. Their achievements were indeed valuable, plus, they were making scientific discoveries:

> A number of stars were found by Messrs. Wright and Palmer to have variable velocities, in other words, they were proven to be attended by a mass of invisible companions. One case discovered by Mr. Wright is of special interest, as it relates to the first-magnitude star *Antares*. Its variable velocity was detected by a comparison of Mills spectrogram secured in Chile with Mills spectrograms of the same star secured earlier on Mount Hamilton.[51]

They were making progress on variable stars and their companions, in this case, on what would become known as the red supergiant star Antares. These discoveries connected scientific tasks in California and Chile. Indeed, the southern observatory seemed to anticipate observations of its northern twin.

[50] Campbell, "Organization and History," 8.

[51] Campbell, "Organization and History," 8–9.

The success of the endeavor so far made it reasonable to assume that the Americans would remain longer. Chileans certainly supposed as much. The popular magazine *ZigZag* described the exploits of the Americans and hinted at the possibilities of a longer stay. "As expected, there will be many chances for the Lick Observatory to agree to hold this branch in Santiago… there are favorable times when the nights are so clear for the observation of the stars that not even the Lick Observatory famous atmosphere can surpass them in its sharpness."[52] The tiny southern observatory, the Lick's twin, became to be more relevant both around the world and in Chile.[53] More than a century later, scholars concluded that the expedition helped initiate the study of astrophysics in Chile.[54]

In 1903, after having carefully documented every detail of the project's developments to Mills in a steady stream of letters, Campbell asked his benefactor for more funding. The outcome was an extension of the project. In 1905 Mills agreed to provide for a further period of observation covering five more years, until 1911, so astronomers included observations of additional starts.[55] But another team would have to continue the mission on San Cristóbal. Heber D. Curtis and his assistant, Mr. Paddock replaced Wright and Palmer in 1905.

3.5 To Be gringo Scientists in 1900s Santiago

Although scientific research was gravitating toward increasingly sophisticated methodologies, interest in connecting scientific inquires and discoveries with social issues persisted. In Western societies, belonging to wealthy circles often entailed some connection to the production of scientific knowledge. Mills may have been thinking about such connections when Campbell asked him to help fund the project. Thus, the director of Lick offered Mills opportunities, even beyond the southern expedition, to invest in science, such as support the Astronomical Society of the

[52] "El observatorio del San Cristóbal," *Revista Zig-Zag*, May 13, 1906.

[53] David S. Evans, *Under Capricorn: A History of Southern Hemisphere Astronomy* (Bristol: IOP Publishing, 1988), 192.

[54] Andreas Reisenegger, "Estrella del pasado," *Revista Universitaria* 116 (2012): 25.

[55] William W. Campbell, "Notes on the D.O. Mills Expedition to the Southern Hemisphere," *Publications of the Astronomical Society of the Pacific* 23, no. 135 (February 1911): 52.

Pacific's idea to publish a "popular" journal.[56] It would be technical, but its pages would include notes and reports that detailed scientific progress on different fronts. The *Journal of Astronomical Society of the Pacific*—as it creatively was named—kept a wide audience aware of what was happening with astronomical research on the Pacific coast and beyond. The *ASP* echoed the connections across the Pacific that Lick Observatory strove to deepen.

The intention of informing a broad public about advances in astronomical research was also present in Chile, where popular magazines such as *ZigZag* and *Pacífico Magazine* published notes about the Observatory. Their coverage often emphasized novelty over science, for initiatives such as the Lick astronomers provoked great curiosity in Chile. The image of the adventurous and eccentric scientist fulfilled the Chileans' expectation.

> Little, very little, they have been seen in the streets of the capital. Rarely they went down the wild slopes of the hill on which the observatory rises, to be confused for a few instants with the rest of the mortals. Embedded in their contemplation of the infinite, for three years they have dedicated all their nights to follow the inscrutable march of the stars trying to penetrate the mysteries of their evolutions and to surprise the supreme rules that direct them.[57]

Articles about the Mills expedition helped to connect the astronomers with the city. "Everyone knows the North American observatory of San Cristóbal," noted *ZigZag*. It is "nothing less than a branch of the great astronomical observatory of Lick of the University of California. Some years ago, the Lick Observatory was actively concerned with developing a vast research program to determine the speed of the brightest stars."[58] Fascination reached into different facets of Santiago society. The American astronomers welcomed interest from the Chilean elite, but were less accommodating of the curiosity of other sectors. With a mix

[56] D.O. Mills Expedition, Correspondence—W.W. Campbell to D.O. Mills. Mount Hamilton, March 9, 1906, UA 36. Ser 04. Box 8 Folder 3, *Lick Observatory Records*, Special Collections, Mary Lea Shane Archives, University of California Santa Cruz.

[57] "El observatorio del San Cristóbal."

[58] "El observatorio del San Cristóbal."

of pride and sorrow, Wright told Campbell that "the observatory itself is looking very well, and its white dome can be seen from all parts of the city, and, unfortunately, has attracted parties of school-children, so that a fence has been erected to keep the curious and destructive out."[59]

Widely circulating periodicals struggled to provide explanations of the scientific techniques that were both accurate and accessible. *ZigZag* explained spectroscopy as the process by which "an apparatus that decomposes the light that reaches us from a certain star determines how quickly that star approaches or moves away from our planetary system." The Chilean journalists did not quite understand the objectives of the expedition. Another passage stated that astronomers were trying to determine "at what speed and in which direction the sun moves with the satellites at infinity."[60] Santiago society perceived these scientist's routines with surprise and amazement:

> It is a curious the life these hermits of science have. They live in the city, at the foot of San Cristóbal; they sleep in the morning, work on their notes and attend the public in their office on Pio IX Street, num. 241, during the afternoon, and go up to the summit at 6 P.M. to spend all night in mystical conversation with the stars until 6 A.M., when they descend again to the plain.[61]

Chileans elaborated a mysterious image of the scientists, almost as if they were characters of science fiction who lived surrounded by Chileans but at the same time far away from the people.

> Their silent life goes unnoticed by the inhabitants of Santiago and they, on the other hand, seem to care very little about the world and its enraged waves. They walk happy, serene, in their fragile *barquilla*, with their eyes set on the heights, in those unusual regions where our limited imagination locates the Invisible Being that directs our worms' crawling on earth. They are happy and deserve all our envy.[62]

[59] D.O. Mills Expedition—Arrangements 1901–1903, September 1, 1903. UA 36. Ser 04. Box 8 Folder 1, *Lick Observatory Records*, Special Collections, Mary Lea Shane Archives, University of California Santa Cruz.

[60] "El observatorio del San Cristóbal."

[61] "El San Cristóbal," *Pacífico Magazine*, November 1914.

[62] "El San Cristóbal."

This sense of mystery, together with the popular press's interpretations of the sidereal problem, did not explain what astronomers were actually researching, but local periodicals still strove to explain the science of the project. In the early twentieth century, that was a valuable intention. It helped connect the American astronomers to local society just as the Pacific Ocean was increasingly connecting the astronomical experiences of Mount Hamilton and San Cristóbal.

Despite the great distance, the twin observatories reflected each other in a game of mirrors. In fact, *ZigZag* declared its hope that the Mills Observatory would be given permanent status in Santiago. This permanence was directly associated with Chile's own development and the dream of continuing with this astronomical task. *ZigZag* stated that astronomers still had many years before they completed their studies and "draw back completely the veil that hides the ethereal life of the planets. Their work will be a precious one for that of another generation of astronomers who will be responsible for collecting the fruits of their efforts and this, in turn, will offer wide field for new research."[63] It was indeed an expectation that Chilean science someday would continue the scientific progress of the American astronomers.

The political, cultural, and scientific relationships between United States and Latin America were growing increasingly closer. Politicians and scientists from Latin America and the United States were organizing Pan-American scientific conferences that aimed to strengthen collaborations across the continent.[64] Of course, these bonds did not ripple through the entire society; in Santiago, they remained in elite spaces of the Chilean oligarchy and were discussed in the ballrooms of *Club de la Unión* and the *Teatro Municipal*. Still, the Chilean press hoped that the American scientific spirit would spread in the south of the world: "we have nothing but congratulations for the arrival of the Yankee [sic] astronomers in Chile". They praised the astronomers' arrival because "They are spirits deeply fond of science, who will not cease to unite the name of our country with that of a great discovery."[65] Scientific progress could prove the Chilean state to be modern, developed, and progressive.

[63] "El observatorio del San Cristóbal."

[64] Rodrigo Ferno, *Science Still Born: The Rise and Impact of the Pan American Scientific Congresses, 1898–1916* (Lincoln: iUniverse, 2003), 28.

[65] "El observatorio del San Cristóbal."

With the expedition remaining in Chile, the number of scientists who traveled south to join the initiative increased. Being far away from California was not easy. Working at night, climbing and descending the hill could be exhausting. Astronomers, therefore, spent a few years in Chile before the team changed. The stay thus combined adventure with permanence dimension. The gringos in Chile began to correspond with their boss in California about aspects of the daily life. Wright described everyday life in Santiago as he passed his time between the city and the hill. "Dr. Palmer and I made daily or nightly climbs to the Observatory. After the first few months a horse was secured to aid in the ascent. This climbing was fatiguing, and after a year or so got to be monotonous, but it was necessary, as the hill is rugged and barren and affords no facilities for family life."[66]

The path uphill was important. They had to make the long climb every day. The varied opinions on the subject underscore the relevance of the daily hike to the observatory. Campbell reported that it took about one hour to reach the observatory, but it could be done on horseback in something over half an hour. He also knew there were two available paths, which could easily be converted into carriage roads, "but the observatory does not need them, I believe."[67] Of course, he was not the one climbing the hill every single day. Constantly going uphill to an observatory was not necessarily part of an astronomer's habit. On Mount Hamilton there were houses and other facilities for astronomers during observations.

Even so, as astronomers mentioned the difficult conditions of the hill, they described Santiago in a different way, almost as if living there was worth the daily hike. For Mr. Wright, "as a place of residence we found Santiago delightful. The social life if the city is as agreeable as the surroundings are beautiful, and many of our happiest recollections are associated to our life there."[68] Wright and his team also described the attitude of Santiago society toward them. "To the great majority of those with whom we came in contact, both Chileans and foreigners, the work

[66] Wright, "Introductory Account," 21.

[67] D.O. Mills Expedition—Arrangements 1901–1903, September 1, 1903. UA 36. Ser 04. Box 8 Folder 1, *Lick Observatory Records*, Special Collections, Mary Lea Shane Archives, University of California Santa Cruz.

[68] Wright, "Introductory Account," 21.

at the Observatory seemed to be a matter of interest, and only rarely did we find any one who did not appear glad to render any needed assistance either to the expedition or to the members personally."[69] This was not by chance. Besides their scientific tasks, life itself was taking place in Santiago.

Just as astronomers filled pages of correspondence with scientific reports, infrastructure requirements, and their impressions about the distant country, the United States Legation in Chile also maintained correspondence with Campbell and thus contributed to the social connections of American astronomers. The Legation Secretary described his experience to Campbell. He had been up to the Observatory on numerous occasions, and had taken friends with him, "and we have been impressed by the amount of good work that has been done by Dr. Wright and Mr. Palmer."[70] Something about the astronomers' work, beyond the scientific tasks themselves, drew the attention of the Secretary. "Both of them have had a lot of the work to do with their own hands which could not be trusted to the workmen of this country, and every time I have been to the top of St. Cristóbal, I have found both the Doctor and Mr. Palmer quite grimy with grease, paint, and rust."[71]

Despite the years of experiences, astronomers had to assume different occupations. At the same time, other misfortunes such as diseases were also part of the astronomers' life thousands of miles away from their trustworthy suppliers of medical attention. Such was the case of the assistant, Palmer, who after enduring a fever was "looking very thin, but as he has a good appetite according to his landlady, I presume he is not exactly in a suffering condition."[72]

[69] Wright, "Introductory Account," 22.

[70] D.O. Mills Expedition—Arrangements 1901–1903, September 1, 1903. UA 36. Ser 04. Box 8 Folder 1, *Lick Observatory Records*, Special Collections, Mary Lea Shane Archives, University of California Santa Cruz.

[71] D.O. Mills Expedition—Arrangements 1901–1903, September 1, 1903. UA 36. Ser 04. Box 8 Folder 1, *Lick Observatory Records*, Special Collections, Mary Lea Shane Archives, University of California Santa Cruz.

[72] D.O. Mills Expedition—Arrangements 1901–1903, September 1, 1903. UA 36. Ser 04. Box 8 Folder 1, *Lick Observatory Records*, Special Collections, Mary Lea Shane Archives, University of California Santa Cruz.

Clearly such a long stay in another country could not be exclusively for work. They could not separate the personal, everyday life from their labor. Therefore, sometimes they described "problems derived from nostalgia." However, it was not only about being homesick, but also that life in a strange place arose some drawbacks. Among them were issues regarding relationship with locals: "The problem of getting Chileans to move rapidly; the rising cost of living; domestic and political disturbances ranging from petty thievery to insurgent revolution; the scarcity of decent housing near the Cerro which constantly had to be climbed; and repeatedly family illnesses."[73]

Nonetheless, American astronomers bonded with the residents of Santiago, even if their stays lasted an average period around three or four years. For this reason, despite some problems, they began to identify themselves with Chile. Paradoxically, when difficulties had been met, endured, or overcame, scientists developed an affection for the place that when the time came for them to leave, they did it with honest unwillingness. Wright, the first astronomer who went to Chile, faced the toughest situation. When describing his experiences, he rarely touched on his personal life. Still, Wright sometimes spoke of their house in Santiago or some trip on horseback to Valdivia in southern Chile, along with his wife and her sister. He also sent photographs of the observatory and of the charming city, but also "The Wrights had much to undergo in the way of riots, bandits, marauders, nearby murders, small-pox scares, and other such deterrent alarms."[74]

Despite these obstacles and difficulties, the astronomers retained a human attachment and even affection for the place that became their home for a few years. Even Wright, who had to overcome these almost dangerous conditions, had to concede that, "From a non-astronomical standpoint, I shall somewhat sorry to leave the country, as I, in common with the rest of the party, have overcome the repugnance which we all felt at first regarding many of the customs and ways of South Americans", and he even concluded, "Look at the best *Chilenos* in the right light, and you will find much to admire and respect."[75] Certainly, this perception was not the first impression the Americans had of the country.

[73] Hard, "Notes on the Directors," 2.

[74] Hard, "Notes on the Directors," 3.

[75] William Wright, quoted in Hard, "Notes on the Directors," 3.

Wright's successor, Heber Curtis, emphasized his disagreement regarding Chile: "Wright's expressed preference for Santiago as a place of residence over any other city on the globe is beyond me, somewhere in the fourth dimension, I guess." Grudgingly, he gave some credit to time: "Perhaps I may get in range of his point of view in five years."[76] Even if he opened a small window, he was categorical when describing his first impression of the country: "I must say that my present opinion of Santiago coincides more nearly with what General Sherman said he would do if he owned both Texas and a Certain Place."[77] Most likely, he was referring to what General Sheridan had said, in the 1860s, "If I owned Texas and Hell, I would rent Texas and live in Hell."

Curtis arrived in Chile in 1906 with his wife, his three children, and his mother-in-law. The large family made it hard to find a place to live and to overcome all the difficulties of arriving in a new country. Still, the arrival was much easier than it had been for his predecessor, for they took over the Wrights' house and bought most of their furniture, so "their domestic settling-in was comparatively easy."[78] Still, Curtis and his family experienced other difficulties. In his correspondence with Campbell, he discussed these issues.

> Constant problems of upkeep, repairs, improvements, accidents, overseeing dilatory laborers, dealing with the owners of the site and with lawyers and with changing government officials, and with the problems of transportation. The slowness of mail, often caused by labor troubles and strikes, was a constant source of anxiety because of delays and damage to supplies shipped from California.[79]

Drawbacks did not end there. There were health issues, which, of course arose also in the United States, but traveling with children in a foreign country added fear and insecurity. It must have been difficult to adjust one's life to a foreign place, considering the differences between California and Chile in the early twentieth century. After a smallpox scare in Chile in June 1906, Curtis had the whole family vaccinated "more than once." Although they were spared from smallpox, his children did

[76] Heber Curtis, quoted in Hard, "Notes on the Directors," 6–7.

[77] Heber Curtis, quoted in Hard, "Notes on the Directors," 7.

[78] Hard, "Notes on the Directors," 4.

[79] Hard, "Notes on the Directors," 4.

get "pneumonia and other infections."[80] Yet, even if at first they seemed unpleased with the host country, a special fondness Chile was born in these astronomers. Curtis recognized in himself the feelings his predecessor had expressed. As he said when expecting Dr. Moore, the scientist who was to replace him in 1909, "We shall leave Santiago with very great regret. Moore finds our attitude toward leaving Chile as difficult to understand as I did Wright's roseate prophecies three years back."[81] Curtis acknowledged how much his emotions and perceptions toward his host country had changed, from quoting Sheridan to feeling sorrow for having to leave.

These astronomers from distant latitudes had ended up on the top of a small hill, in the suburbs of a city at the end of the world. It was a crossroads of faraway experiences and surprising similarities. The reflection of their lives on Mount Hamilton acquired new strength in the south. The effort to unveil those fundamental questions about our universe made it worth the transfer, the distances, the unexpected and the novelties that this new world brought. It was the same continent, which was reflected from north to south. At the same time, their stay opened possibilities to resolving the sidereal problem that kept them awake at night in the darkness of San Cristóbal Hill.

3.6 A Shaky Observatory: The 1906 Earthquake

An unexpected event deeply relevant to the astronomers' actions in the south occurred on August 16, 1906. At 8 p.m., an 8.6 earthquake roared through the country. The situation was critical, with more than 2300 deaths and 20,000 injured, and the collective hysteria, fear, and uncertainty.[82] The devastation was most intense near the epicenter in Valparaíso. But the earthquake imperiled the astronomical equipment, the observatory facilities, and the astronomers themselves in Santiago only 110 kilometers away. For the American astronomers, surviving the earthquake was among the most arresting moments of their stay in Chile. The disruption of communication did not help the anxiety both in

[80] Heber Curtis, quoted in Hard, "Notes on the Directors," 4.

[81] Heber Curtis, quoted in Hard, "Notes on the Directors," 7.

[82] Alfredo Rodríguez and Carlos Gajardo, *La catástrofe del 16 de agosto de 1906 en la República de Chile* (Santiago: Barcelona, 1906).

California and in Santiago. When informing about the quake, Campbell told Mills mails were not dispatched from Santiago for two weeks, so he only got Curtis's letter more than two months after the earthquake.[83] Still, he was lucky enough to receive some news about one week after the tremors; Campbell finally read the telegram with relief: "Observatory at Santiago, Chile, undamaged and all our people are well."[84]

For Curtis, this event recalled recent experience in California. Shortly after Dr. Curtis arrived to Chile, he heard "the terrible and incomplete news from San Francisco in April about the earthquake and fire."[85] Four months separated the two seismic events of similar magnitude, which generated more feelings of strange familiarity, now in terms of chaos and fear. Chile and California, it seemed, connected and reflected each other, much like the mirrors of the Mills spectrograph. The 1906 earthquake in Valparaíso caused fires and riots.

It was impossible not to recall the situation in San Francisco just a few months before. Nonetheless, the Chilean political and administrative organization was significantly different from that of the United States. The president of the so-called Parliamentarian Republic, Germán Riesco, declared Valparaíso in state of constitutional exception, a legal figure used for emergencies. It temporarily suspended the constitutional procedures so that the government could efficiently mobilize resources and personnel. But it also left room for excesses. The president appointed Army Major Luis Gómez Carreño in charge of the city. One of Luis Gómez Carreño's prerogatives was to apply martial law to those caught stealing or looting.[86]

When they finally managed to reestablish communication, Curtis told every detail of how they had survived the quake. "We had a very severe earthquake last night at about eight o'clock. There were two hard shocks a few seconds apart and lasting about four minutes. I started counting the seconds to time it, changed to a sudden wish to see the lamps

[83] D.O. Mills Expedition, Correspondence—W.W. Campbell to D.O. Mills. Mount Hamilton, October 27, 1906. UA 36. Ser 04. Box 8 Folder 3, *Lick Observatory Records*, Special Collections, Mary Lea Shane Archives, University of California Santa Cruz.

[84] D.O. Mills Expedition, Correspondence—W.W. Campbell to D.O. Mills (telegram). Mount Hamilton, August 24, 1906. UA 36. Ser 04. Box 8 Folder 3, *Lick Observatory Records*, Special Collections, Mary Lea Shane Archives, University of California Santa Cruz.

[85] Hard, "Notes on the Directors," 4.

[86] *Terremotos en Chile. Valparaíso, Chillán, Valdivia* (Santiago: Museo Histórico Nacional de Chile, 2009), 30.

put out, and ended in a rush for the rain-drenched patio."[87] That rainy evening in August had generated such a chaos in the city that, as Curtis described, "the din was very great and all around could be heard the shrieks of the natives."[88] Something that surprised Curtis was that in the southern country, there were two different words for a seismic event, depending on its magnitude: "I send you some papers describing what all agree was a 'terremoto' and not a 'temblor.'"[89] Earthquakes in Chile were so common, it was embedded in their own identity,[90] and the locals differentiated between them. The latter meant just tremors, of almost no importance, while the former referred to an intense tectonic movement that caused greater damage.

Instinctively, amid the confusion and aftershocks, Curtis and his family struggled to understand what had happened. The damage was extensive: "a corner of the garden wall down, the rear wall of the house nearly all down on our house and entirely so on three other houses in the block, a five-inch crack at the ceiling where the side wall of the dining room had nearly gone down besides fallen plaster and stucco everywhere."[91] Campbell sent Mills a report on the situation from the acting astronomer in Santiago. He was deeply worried by Curtis and his family, sorrowed that his house was severely damaged but glad they were uninjured, nor had to sleep outside in the cold rain, as most of Santiago's inhabitants did.[92] Still, Campbell was concerned about the damages to Curtis's house and requested Mills approval for aid.[93] The earthquake's magnitude, as well as the intensity and frequency of the aftershocks, explained the conditions of the astronomer's house. Never abandoning his scientific spirit, Curtis detailed that terrifying evening:

[87] Heber D. Curtis, quoted in Hard, "Notes on the Directors," 4.

[88] Heber D. Curtis, quoted in Hard, "Notes on the Directors," 4.

[89] Heber D. Curtis, quoted in Hard, "Notes on the Directors," 4.

[90] Bárbara Silva and Alfredo Riquelme, *Una identidad terremoteada. Comunidad y territorio en el Chile de 1960* (Santiago: Ediciones Universidad Alberto Hurtado, 2018).

[91] Heber D. Curtis, quoted in Hard, "Notes on the Directors," 4.

[92] "Nearly all Santiago slept out doors that night, in spite of the rain. Santiago was pretty badly scared and slept in coaches or in the street cars [sic] or without protection." Heber D. Curtis, quoted in Hard, "Notes on the Directors," 5.

[93] D.O. Mills Expedition, Correspondence—W.W. Campbell to D.O. Mills. Mount Hamilton, October 27, 1906. UA 36. Ser 04. Box 8 Folder 3, *Lick Observatory Records*, Special Collections, Mary Lea Shane Archives, University of California Santa Cruz.

> I regret that I did not keep an accurate record of all the separate shocks, but the earth was literally in continuous movement for four hours after the shock. During all this time the parlor lamp kept quivering, with only an occasional intermission of rarely over thirty seconds. Every once in a while came a heavier tremble that sent us hurrying to the doors.[94]

The earthquake had consequences for the Curtis family life. His daughter Margaret was extremely frightened by the quake. Curtis could not finish writing about his impressions without being interruption by further shocks. "When I trimmed down the light and got ready to take Margaret out if it got worse. Poor child, she is in a pretty bad state nervously over the shake, and frequently expresses her desire to leave this country."[95] This passage was not part of the scientific report but a comment on a dramatically human event that Curtis lived that evening.

After ensuring that his family was safe, and that the house was not going to collapse, Curtis turned to the observatory. Fortunately, he remembered the exact position in which he had left the telescope the day before. His first desire was to go at once to the summit, but he was heavily out-voted. He remembered the tube of the telescope was left standing straight up, and the big mirror had been placed where it would be practically impossible for the falling tube to hit it. The concern was that "the good old prisms might have been broken, or the secondary fallen off the place I had left it."[96] The relief when climbing the hill, reaching the summit, and seeing what had happened at the observatory was huge. Campbell must have felt the same relief in California.

With even greater relief and even pleased, he informed Mills the Observatory in Chile suffered no damage: "The total loss on the summit would scarcely exceed $5,00 [sic] and relates to dark-room photographic apparatus of glass thrown from the shelves."[97] This was good news and the astronomers were grateful. The risk that such an intense earthquake posed for the delicate equipment of a spectrographic observatory was

[94] Heber Curtis, quoted in Hard, "Notes on the Directors," 5.

[95] Heber Curtis, quoted in Hard, "Notes on the Directors," 5.

[96] Heber Curtis, quoted in Hard, "Notes on the Directors," 5.

[97] D.O. Mills Expedition, Correspondence—W.W. Campbell to D.O. Mills. Mount Hamilton, October 27, 1906. UA 36. Ser 04. Box 8 Folder 3, *Lick Observatory Records*, Special Collections, Mary Lea Shane Archives, University of California Santa Cruz.

considerable. Some of the retaining walls as well as nearby buildings were damaged but not beyond repair.[98] The greatest concern was the spectrograph's equipment, and of course, the mirrors, for which there were no replacements in Chile. For Campbell, it was a happy coincidence that Curtis worked in the nearby rooms; most importantly, he had almost completed his work on the telescope and mirrors before the quake. He told Mills that the additional small room Curtis was building and other improvements to the telescope and spectroscope were practically completed before the quake. This was indeed most fortunate, for it would have been almost impossible for Curtis to get help and supplies after the disaster.[99]

The road that ascended the hill was another matter; it was "badly damaged through the upper bank caving into the road in some places and the lower bank slipping out of the road on other ones." Being able to get to the observatory was very important and after the quake the road was filled with rocks which have come down from the quarry. In addition, there were two horses that the falling rocks had killed, but the electric line stood.[100] Given the critical issue of having an appropriate road to go uphill, this was a matter of concern for Campbell. It was very likely that the Chileans would focus on greater emergencies caused by the quake than a desolated road to go to the summit of a rural hill. Mills wrote that Wright would "have to make some repairs" on his own.[101]

An unexpected reaction by Santiago inhabitants added to the uncertainty. Some hoped that Curtis could impart scientific wisdom on earthquakes, and most of all, on predicting them. Chileans believed the American scientists had superior knowledge about seismic events, and it spread among people of all types, status, and occupations.

[98] Hard, "Notes on the Directors," 5.

[99] D.O. Mills Expedition, Correspondence—W.W. Campbell to D.O. Mills. Mount Hamilton, October 27, 1906. UA 36. Ser 04. Box 8 Folder 3, *Lick Observatory Records*, Special Collections, Mary Lea Shane Archives, University of California Santa Cruz.

[100] Heber Curtis, quoted in Hard, "Notes on the Directors," 6.

[101] D.O. Mills Expedition, Correspondence—W.W. Campbell to D.O. Mills. Mount Hamilton, October 27, 1906. UA 36. Ser 04. Box 8 Folder 3, *Lick Observatory Records*, Special Collections, Mary Lea Shane Archives, University of California Santa Cruz.

> Since the shock, fame has been suddenly thrust upon me. All day long people have been at my door, now *roto* women, another time army officers, policemen from this or that *Comisaria*, and two or three detachments from the chief himself; as I walk along the streets in this part of the city I am stopped again and again by children sent by their parents to interrogate me.[102]

Without a doubt, this drew the astronomers' attention since Campbell related the episode to Mills. "For many days following the earthquake," Campbell wrote his benefactor that Curtis was "constantly in demand to predict additional earthquakes that were about to come. Even the President of Chile dispatched messengers with mounted escorts to ask him whether there would be any more earthquakes this evening, and so on."[103] For Curtis, this was the highlight of the situation.

> The climax was reached tonight when a brilliant aide-de-camp with a small lancer escort came bearing a note from His Excellency the President of the Republic. And they all had only one question to ask, 'If you please, will there be another *terremoto* tonight?' I assured them all that there would doubtless be many slight and un-important shocks tonight and tomorrow, but in all probability none sufficient to cause alarm. I hope I may prove to have been a good prophet, though I guess that that last shock a few minutes ago may have shaken the faith of some.[104]

The fact that neither he nor anyone could predict a quake struck Curtis as a bit comical. Perhaps due to the frequency Chileans experienced seismic events, or maybe because of ignorance, the scientific relations promoted by the quake were of great interest and surprise for a scientist like Curtis. "The funniest part of it, from the scientific standpoint, is the corroboration which the Chilean who works at the

[102] Heber Curtis, quoted in Hard, "Notes on the Directors," 6.

[103] D.O. Mills Expedition, Correspondence—W.W. Campbell to D.O. Mills. Mount Hamilton, October 27, 1906. UA 36. Ser 04. Box 8 Folder 3, *Lick Observatory Records*, Special Collections, Mary Lea Shane Archives, University of California Santa Cruz.

[104] Heber Curtis, quoted in Hard, "Notes on the Directors," 6.

Weather bureau has got out of it. The Moon still plays an important part in weather prediction here (I wish I had some prognostications to send you)."[105] Then, he mentioned that "atmospheric or seismic disturbances" were predicted for that day, due to the Moon and Jupiter conjunction.

Curtis was probably referring to the predictions of Capitan Middleton of the Chilean Army. Just a few days before, citing predictions from the Meteorology Section of the Department of the Maritime Territory, *El Mercurio de Valparaíso* announced that there would soon occur a "conjunction of Neptune with the Moon." As a result, the location of the stars posed danger for Valparaíso.[106]

Repeatedly printed and evoked, this prediction was a sort of fantasy that had nothing to do with the remote ability to foresee when and where these seismic events would occur.[107] Still, the quake's impact on Chilean society was real and turning to the astronomers became almost habit. The following year, on the anniversary of the earthquake, Dr. Curtis received again two policemen, sent to him as the authority wanted to know "whether the populace might expect a recurrence on the anniversary of the disaster."[108] Even if nothing happened in that anniversary, given the country's seismic condition the Chilean government took some actions. In 1907 the government hired the French seismologist, Earl Fernand Montessus de Ballore, as Chief of the Seismologic Observation Service. His task was to report about quakes and train a group of Chilean scientists, specialized in quakes, and systematize studies on earthquakes.[109]

[105] Heber Curtis, quoted in Hard, "Notes on the Directors," 6.

[106] *El Mercurio de Valparaíso*, August 6, 1906.

[107] Jose Luis Giordano, *La predicción del terremoto de 1906 ¿ciencia o fantasía? Una aproximación a la historia perdida bajo la leyenda del Capitán Middleton* (Riga: Académica Española, 2013).

[108] Hard, "Notes on the Directors," 6.

[109] Rosa Urrutia and Carlos Lanza, *Catástrofes en Chile 1541–1992* (Santiago: La Noria, 1993), 164; *Terremotos en Chile*, 32.

Surviving that shaky year of 1906, in the north and the south, was a source of peace and happiness for the director of the Lick Observatory. Campbell communicated exactly this feeling to Mills. "It is a matter of congratulation that the Lick Observatory on Mount Hamilton and the Mills Observatory on San Cristóbal, Santiago, passed safely through the two great earthquakes."[110] Mr. Mills had suffered quite a loss, and his elegant corporate building in San Francisco was reduced to ashes. At least his scientific investments, in California as well as in the Southern Hemisphere, were saved from significant damages. This was quite a surprise. "There is no doubt that the Chile earthquake was more severe than our California one, and we have abundant causes for thanksgiving that the two observatories passed safely through the two earthquakes."[111] The two observatories were still standing without major losses.

The next year, Campbell still informed on the consequences of the quake, when the emergency had already long passed. "Everything is going well in Chile, except that prices have advanced about fifty per cent since the earthquake. Fortunately, our instruments at the Observatory were nearly all completed before the catastrophe, but Dr. Curtis is quite seriously affected by the price of living, as his family numbers six members."[112] Overall, work was back to normal at San Cristóbal, and scientific progress continued. After strikes, protests, riots, epidemics, and vaccinations, the gringo scientists had almost adapted to the local life.

[110] D.O. Mills Expedition, Correspondence—W.W. Campbell to D.O. Mills. Mount Hamilton, August 24, 1906, UA 36. Ser 04. Box 8 Folder 3, *Lick Observatory Records*, Special Collections, Mary Lea Shane Archives, University of California Santa Cruz.

[111] D.O. Mills Expedition, Correspondence—W.W. Campbell to D.O. Mills. Mount Hamilton, October 27, 1906. UA 36. Ser 04. Box 8 Folder 3, *Lick Observatory Records*, Special Collections, Mary Lea Shane Archives, University of California Santa Cruz.

[112] D.O. Mills Expedition, Correspondence—W.W. Campbell to D.O. Mills. Mount Hamilton, January 31, 1907, UA 36. Ser 04. Box 8 Folder 3, *Lick Observatory Records*, Special Collections, Mary Lea Shane Archives, University of California Santa Cruz.

Bibliography

American Association of the Advancement of Science. "The D.O. Mills Astronomical Expedition," *Science* 17, New Series, no. 428 (March 1903): 436.

Bigg, Charlotte, David Aubin, and Philipp Felsch. "Introduction: The Laboratory of Nature—Science in the Mountains," *Science in Context* 22, no. 3 (2009): 311–321.

Campbell, William W. "Organization and History of the D.O. Mills Expedition to the Southern Hemisphere," *D.O. Mills Expedition Cerro San Cristóbal, Santiago, Chile from Its Organization in December 1900 to March 1906*, ed. W.W. Campbell and W.H. Wright, Vol. IX, 5–14. Sacramento: W. W. Shannon Superintendent State Printing, 1907.

Campbell, William W. "Notes on the D.O. Mills Expedition to the Southern Hemisphere" *Publications of the Astronomical Society of the Pacific* 23, no. 135 (February 1911): 52–54.

D. O. Mills Expedition, UA 36. Ser 04. Boxes 1 to 12 *Lick Observatory Records*, Special Collections, Mary Lea Shane Archives, University of California Santa Cruz.

El Mercurio de Valparaíso, August 6, 1906.

"El observatorio del San Cristóbal," *Revista Zig-Zag*, May 13, 1906.

"El San Cristóbal" *Pacífico Magazine*, November 1914.

Evans, David S. *Under Capricorn: A History of Southern Hemisphere Astronomy.* Bristol: IOP Publishing, 1988.

Ferno, Rodrigo. *Science Still Born: The Rise and Impact of the Pan American Scientific Congresses, 1898–1916.* Lincoln: iUniverse, 2003.

Garcés, Mario. *Crisis social y motines populares en el 1900.* Santiago: Documentas, 1991.

Giordano, Jose Luis. *La predicción del terremoto de 1906 ¿ciencia o fantasía? Una aproximación a la historia perdida bajo la leyenda del Capitán Middleton.* Riga: Académica Española, 2013.

Hard, Beverly. "Notes on the Directors of the D.O Mills Expedition to Chile," 1972. Expeditions, UA 36. Ser 04. Box 12 Folder 6, *Lick Observatory Records*, Special Collections, Mary Lea Shane Archives, University of California Santa Cruz.

Leiva, Ricardo. "Atacama: a 100 años del informe Curtis," *Revista de Humanidades* 3 (2010): 10–22.

Morris, James. *Las elites, los intelectuales y el consenso.* Santiago: Pacifico, 1967.

Reisenegger, Andreas. "Estrella del pasado," *Revista Universitaria* 116 (2012): 24–27.

Rodríguez, Alfredo and Carlos Gajardo. *La catástrofe del 16 de agosto de 1906 en la República de Chile.* Santiago: Barcelona, 1906.

Silva, Bárbara. *Identidad y nación. Patria Vieja, Centenario y Bicentenario.* Santiago: Lom, 2008.

Silva, Bárbara and Josefina Cabrera. *Chile, 100 días en la historia del siglo XX.* Santiago: Planeta, 2015.

Silva, Bárbara and Alfredo Riquelme. *Una identidad terremoteada. Comunidad y territorio en el Chile de 1960.* Santiago: Ediciones Universidad Alberto Hurtado, 2018.

Terremotos en Chile. Valparaíso, Chillán, Valdivia. Santiago: Museo Histórico Nacional de Chile, 2009.

Urrutia, Rosa and Carlos Lanza. *Catástrofes en Chile 1541–1992.* Santiago: La Noria, 1993.

Wright, William H. "Introductory Account of the D.O. Mills Expedition," *D.O. Mills Expedition Cerro San Cristóbal, Santiago, Chile from Its Organization in December 1900 to March 1906*, ed. W.W. Campbell and W.H. Wright, Vol. IX, 15–24. Sacramento: W. W. Shannon Superintendent State Printing, 1907.

CHAPTER 4

The Americans Want to Stay

Abstract This chapter examines how the Expedition remained longer than initially expected in Chile, and describes their progress in astronomical research. While explaining exactly what was the sidereal problem, we can understand the avant-garde questions they were posing. They were getting to know radial velocities, to analyze them, and to study binary stars. In their daily lives, whereas they measured the stars spectra, the astronomers were still constantly surprised with the odd habits of Chileans. They also decided to explore other areas of the country, and had an amazing adventure in the Atacama Desert, which decades later occupied a fundamental place in global astronomy. This chapter connects the adventure of the Lick Observatory astronomers with the current development of cutting edge, international astronomy in the Atacama region.

Keywords Sidereal problem · Binary stars · Chilean society · Atacama

The initial arrival to Valparaíso in 1903 seemed increasingly distant. The years went by and the Mills Expedition became more stable in Chile. The challenges of getting settled lay behind and the project was functioning smoothly and efficiently. Astronomers came and went every three or four years, but the shock of arriving in Chile was increasingly gentle. A tone of ordinariness became part of the travelers' stories. Traumatic, or simply troubling, events such as the 1906 earthquake were sliding into the

B. K. Silva, *Astronomy at the Turn of the Twentieth Century in Chile and the United States*, Palgrave Studies in the History of Science and Technology, https://doi.org/10.1007/978-3-030-17712-6_4

provinces of memory. Instead, the daily climb up the hill and relationships with Chilean society formed the bulk of the scientists' lives. Some even began to understand that remote place of the Americas.

The expedition was enduring longer than expected and hopes of extending the time even further grew. There were some changes, some vertiginous, other almost imperceptible, that affected the work of the Lick astronomers in Chile. What remained constant was the close connection between Chile and California under Campbell's leadership—and, of course, the goal of resolving the sidereal problem, now in the second stage of the project.

4.1 The Sidereal Problem from the South

"The work of the expedition will be largely technical," Campbell explained, yet its purpose was to contribute to the solution of a perfectly definite problem. What had compelled the scientists to travel thousands of miles, settle in the southern hemisphere, and invest thousands of dollars was the so-called sidereal problem. At state was a new understanding of the dynamics of the galaxy: "It is well known that the observed positions and motions of celestial objects are influenced not only by the motions of the objects themselves, but also by motions of the observer."[1] At the time, to understand the motion of celestial objects, three components were at work.

The first one was the earth's rotation on its own axis, which was well known and could be cleared from observation tasks. The second, the Earth's translational movement around the sun, was also known, even if there was less than 1% of uncertainty on its orbit. Despite this, the astronomers trusted that the margin of error was about to be solved. The third motion component was the key to this expedition: the movement of the solar system through space. "The elements of this motion are not well known. In fact, a general knowledge of them constitutes one of the most pressing problems in astronomy, and it is to contribute to the solution of this problem that the Mills expedition to the southern hemisphere has been organized."[2]

[1] William W. Campbell, "Will Determine Whether Solar System Speeds in Its Flight Through Space," D.O. Mills Expedition, February 23, 1903, UA 36. Ser 04. Box 8 Folder 1, *Lick Observatory Records*, Special Collections, Mary Lea Shane Archives, University of California Santa Cruz.

[2] Campbell, "Will Determine Whether".

Campbell's hypothesis was that not only our planet or the planets in our solar system were in motion, but also that all celestial bodies were moving. He argued that for quite some time astronomers' studies had focused on understanding our solar system. But, "heretofore, the energies of astronomers have been devoted largely to problems connected with our Solar System—our sun, its planets, their moons, and the comets—the motions of the bodies in the system, the motion of the system, the forms, the dimensions, the physical conditions, and the history of its members." These questions, therefore, were related to what Campbell called the "solar problem."[3] Now, as scientific knowledge and technology evolved, it became possible to reach distant areas of the galaxy. Yet, instead of resolving questions regarding the solar system, scientific progress only multiplied them. One of Campbell's concerns was to understand what was happening in those distant spaces, not only in the solar system, but also in the stellar system.

> Improvements in instruments, methods and financial resources are enabling us to study the incomparably more remote stars, nebulae and other bodies composing the stellar system; we want to know how the stars and nebulae are arranged and distributed in the space, how they are related to each, how they are moving, what their physical conditions are and what their histories have been.[4]

The "sidereal problem" therefore constituted a step further toward the understanding of the accessible universe. It was a step that defied the boundaries of astronomy. To the director of the Lick Observatory, this was *the* big astronomical problem—and he wanted to solve it. In the "universality of its interest and in the grandeur of its scope," the sidereal problem surpassed all the other "physical inquiries which engage

[3]D.O. Mills Expedition, Correspondence—W.W. Campbell to D.O Mills. April 4, 1905. UA 36. Ser 04. Box 8 Folder 3, *Lick Observatory Records*, Special Collections, Mary Lea Shane Archives, University of California Santa Cruz.

[4]D.O. Mills Expedition, Correspondence—W.W. Campbell to D.O Mills. April 4, 1905. UA 36. Ser 04. Box 8 Folder 3, *Lick Observatory Records*, Special Collections, Mary Lea Shane Archives, University of California Santa Cruz.

the attention of scientific men'."[5] His conviction that the Lick expedition could obtain critical information was part of his persuasive correspondence with his benefactor and his colleagues. His scientific spirit led him to declare that, "The solution of this tremendous problem calls for greater labor and devotion, but the solution will certainly be made. The human race will someday know the answers to all these questions."[6] For Campbell, attacking the sidereal problem entailed resolving questions such as whether the stars' distribution was uniform in volume, whether their motion was random in every possible direction or if it followed a certain yet still unknown pattern.

To answer these questions, he needed information from spectroscopic measurements of radial velocities of many stars. In order to show that his argument was not merely a product of his own theories, Campbell drew on the Carnegie reports. Clearly, other scientific institutions, not directly linked to the University of California, also supported his ideas. The report corroborated the need to make these measurements—and especially on stars that were not visible from the northern hemisphere. He thus validated the purpose of the Lick Observatory's work on the southern skies.

Observations of the motion of celestial bodies in relation to the observer's location from the northern hemisphere alone excluded a huge portion of the sky. To extrapolate from such limited data would be to speculate on stellar movements that astronomers were just beginning to understand. Thus, the Carnegie report assigned a remarkable responsibility to the scientific expedition in Chile. The goal was to have information on the features and movements of all stars in the sky, brighter than magnitude 5.[7] Campbell's scope was precisely to secure the measurement of radial velocities of all southern stars that he considered to be critical

[5] D.O. Mills Expedition, Correspondence—W.W. Campbell to D.O Mills. *Carnegie Institution Year Book for 1903*, page 24, April 4, 1905. UA 36. Ser 04. Box 8 Folder 3, *Lick Observatory Records*, Special Collections, Mary Lea Shane Archives, University of California Santa Cruz.

[6] D.O. Mills Expedition, Correspondence—W.W. Campbell to D.O Mills. April 4, 1905. UA 36. Ser 04. Box 8 Folder 3, *Lick Observatory Records*, Special Collections, Mary Lea Shane Archives, University of California Santa Cruz.

[7] D.O. Mills Expedition, Correspondence—W.W. Campbell to D.O Mills. *Carnegie Institution Year Book for 1903*, page 38, April 4, 1905. UA 36. Ser 04. Box 8 Folder 3, *Lick Observatory Records*, Special Collections, Mary Lea Shane Archives, University of California Santa Cruz.

to the sidereal problem. In the end, what he pursued was nothing less than an understanding of the structure of the galaxy. This research would finally uncover those virgin skies of the south that remained in the dark despite their stellar bodies' light. Campbell's goal to explore that portion of the sky—to measure and someday to understand it—had become his life's purpose.

The work was not simple, and every step was arduous. Yet, every plate on which the astronomers registered a star's spectrum helped chip away at the mysteries, gradually yielding more precise knowledge of the brighter stars. After fifteen months of work with the observatory fully functioning, the Mills Expedition had discovered ten binary stars in the southern skies.[8] Such findings circulated beyond scientific circles, into major popular periodicals such as the *New York Times* that celebrated the "Discovery of 20 double stars" and declared that the "Expedition sent out by D.O. Mills is meeting success."[9] Certainly, the pathway of discoveries was expected continue, but only if the project had more time.

4.2 "We Need More Time"

Since the beginning, Campbell likely intended to find a way to extend the expedition's stay in Chile, beyond the initially planned two or three years. Proposing a longer stay for the southern expedition from the beginning was probably not a good idea. Mills had already been reluctant to provide any funding at all, so a short time frame was a more probable proposition. Still, Campbell got the money to lengthen the astronomers' mission in Chile as soon as 1905. The underlying argument was of course the importance of the expedition for the sidereal problem, and the promising outcomes they were already collecting. In 1905, Mr. Mills agreed to provide for the expedition's support through a

[8] "Every plate makes a distinct contribution to the problem of Solar Motion, for which the expedition was organized; and in addition, as by-products, five spectroscopic binary stars have been discovered." D.O. Mills Expedition. Correspondence—W. W. Campbell to D. O Mills. January 21, 1905, UA 36. Ser 04. Box 8 Folder 3, *Lick Observatory Records*, Special Collections, Mary Lea Shane Archives, University of California Santa Cruz.

[9] *New York Times*, March 24, 1905.

further period of five years, including all the improvements in and additions to the instrumental equipment.[10]

Mills was probably not ecstatic about spending lots of money to extend the expedition. Yet since he had already made the sizeable initial investment, it made sense to keep the astronomers working in the south. The possibilities of huge discoveries, associated with his name, loomed. But as Campbell said, a longer stay in Chile would require improvements to the observatory's facilities. In each report and presentation regarding such improvements, Campbell's team was careful to mention that Mills's sponsorship had made them possible.[11] This second stage had the goal of "securing a greatly increased number of velocities of southern stars covering a larger range of magnitude chiefly for their values in studies relating to the structure of the sidereal universe."[12]

For this next stage, Wright, who had completed the difficult tasks of transferring and installing the scientific equipment, returned to the Lick Observatory in California. Lick colleague Heber D. Curtis arrived in Chile in March 1906 to replace Wright as acting astronomer of the southern station. At the same time, George H. Paddock, from the University of Virginia, replaced Palmer as assistant.[13]

The first task for Curtis was to carry out the improvements Wright had deemed indispensable as soon as possible. As the objective of the Mills expedition enlarged, its facilities also grew; they began the construction of a small machine room with dormitories for observers.[14] Finally, the astronomers would have a place to rest up there at the summit. From then on, changes in acting astronomers and their assistants became usual. This contributed continuity to the research and strengthened the relationship between Chile and California.

Even with these improvements, the daily life of the astronomers in the south remained arduous. The solitude and the daily climb were no less

[10]William W. Campbell, "Organization and History of the D.O. Mills Expedition to the Southern Hemisphere," *D.O. Mills Expedition. Cerro San Cristóbal, Santiago, Chile. From Its Organization in December 1900 to March 1906*, ed. W.W. Campbell and W.H. Wright, Vol. IX (Sacramento: W. W. Shannon Superintendent State Printing, 1907), 11.

[11]Heber D. Curtis, "Recent Changes at the Observatory of the D.O. Mills Expedition," *Publications of the Astronomical Society of the Pacific* 19, no. 116 (October 1907): 227–233.

[12]Campbell, "Organization and History," 12.

[13]Campbell, "Organization and History," 12.

[14]"El observatorio del San Cristóbal," *Revista Zig-Zag*, May 13, 1906.

taxing. The delicate task of measuring the unknown southern stars' radial velocities could hardly be entrusted to scientists who were exhausted. Therefore, new changes in the astronomical team came three years later, in 1909, when Dr. Joseph H. Moore arrived in Santiago, and once again connections with Chilean society proved invaluable.[15]

The new assistant was Roscoe F. Sanford, who would work with Moore, his wife (and assistant) Federica Chase, and Paddock (who stayed a while longer after his appointment). Sanford's specialization was the analysis of southern stars, or more specifically, stellar spectroscopy and study of the radial velocities of southern planetary nebulae. He worked on velocities of the Magellanic Clouds, which marked one of the first successful attempts to measure the velocities of objects outside the Galaxy.[16] This was a turning point in astronomy. Scientists could now reach what was invisible to the naked eye and they could go further, into extragalactic bodies. Hence, following the initial period between 1903 and 1905 and the five-year extension to 1911, Campbell obtained Mills commitment of funding another two years.[17]

4.3 Mills Expedition's Awards: Technology, Publications, and Prizes

Recognition from different points of view could strengthen the Mills expedition and attract funds. In a world that was beginning to integrate changing and even vertiginous technology, such acknowledgment was fundamental. Whatever technological innovations the astronomers accomplished would also increase the recognition from their peers. Campbell emphasized this point in his correspondence. For him, the lack of observatories in the Southern Hemisphere resulted in a strong demand for observations on the southern stars, hence their mission was

[15] Hebert D. Curtis, "Recent Progress in the Work of the D.O. Mills Expedition to the Southern Hemisphere," *Publications of the Astronomical Society of the Pacific* 21, no. 128 (October 1909): 201.

[16] E. Wilson, "Roscoe Frank Sanford, 1883–1958," *Publications of the Astronomical Society of the Pacific* 70, no. 415 (1958): 360.

[17] William W. Campbell, "Notes on the D.O. Mills Expedition to the Southern Hemisphere" *Publications of the Astronomical Society of the Pacific* 23, no. 135 (February 1911): 52–53.

"recognized by all astronomers and frequently commented upon."[18] Not only their observations but also the cutting-edge technological capacity installed in the Mills Expedition's observatory was recognized and detailed, even in Chile. *ZigZag* magazine explained these characteristics:

> The main equipment of the facilities consists of a large reflecting telescope, one of the most powerful telescopes that are currently in service in the entire world. There are only one or two superior [telescopes] under construction. The large mirror is 37 inches in diameter, 6 inches thick and weighs five hundred pounds. It is considered one of the most difficult tasks in the world to coat such a large crystal disk with the silver layer necessary to give it a good reflecting surface.[19]

The technology did not end with the mirror or the reflector. Additional equipment essential for spectroscopy also included a cooling apparatus to maintain the great mirror's temperature.[20] These innovations in instruments and techniques were also the subject of publications in the scientific community of the United States that detailed the equipment, methods, and instruments that made possible the observations in the southern station of the Lick Observatory.[21] These publications on the technical innovation and the research in Chile validated their work.

Of course, it was about more than publishing. Their benefactor Mills had to be informed of everything. In addition to highlighting the astronomers' accomplishments, publications celebrated Mills as a patron of international scientific networks. Campbell never missed an opportunity to tout the expedition's accomplishments, as he did when mailed nearly 700 copies of the Volume IX, Publications of the Lick Observatory, which contained an account of the organization of the D.O. Mills

[18] D.O. Mills Expedition, Correspondence—W.W. Campbell to D. O Mills. January 21, 1905. UA 36. Ser 04. Box 8 Folder 3, *Lick Observatory Records*, Special Collections, Mary Lea Shane Archives, University of California Santa Cruz.

[19] "El observatorio del San Cristóbal."

[20] D.O. Mills Expedition, Correspondence—W.W. Campbell to D.O Mills, September 27, 1907 UA 36. Ser 04. Box 8 Folder 3, *Lick Observatory Records*, Special Collections, Mary Lea Shane Archives, University of California Santa Cruz.

[21] Heber D. Curtis. "Temperature Control for Silvered Specula," *Astrophysical Journal* 26 (November 1907).

Expedition and description of its instruments and methods.[22] At the same time, astronomers and physicists who did not have a direct relation with the expedition reviewed these publications in other scientific journals.[23]

As necessary as it was to ensure scientific accuracy and assertiveness, the astronomers also had to remained engaged with both the American and international scientific communities. Hence, talking about some of expedition's findings, Campbell explained that, "The paper contains many very valuable results of the experiences of the Expedition, but does not aim to publish as yet the scientific results which the Expedition has been working for."[24] Considering that the project was work in progress, the amount and pace of information to publish remained uncertain. But they had to publicize at least some data that showed they were making progress. In this regard, numbers were relevant: "As stated on page 8 of the paper, the observation list was made up of 145 stars in the third of the sky that is around the south Pole, after excluding 92 stars whose spectra were photographed and found not to contain lines suitable for accurate measurement."[25] Such details conferred credibility and gravitas to the astronomers' research. Plus, Campbell needed to validate his leadership on this scientific enterprise as a trustworthy planner; he was proud to state that the number of plates secured during the two years of observation by Wright and Palmer were almost exactly those called for in his original plan.

The accuracy of the work the Lick Director had planned and designed had to correlate with the quality of the research they were developing in Chile, and at the same time to show they had much more to explore. For

[22] D.O. Mills Expedition, Correspondence—W.W. Campbell to D.O Mills, September 27, 1907 UA 36. Ser 04. Box 8 Folder 3, *Lick Observatory Records*, Special Collections, Mary Lea Shane Archives, University of California Santa Cruz.

[23] For example, Edwin B. Frost, "Radial Velocities of 150 Stars South of Declination—20° Determined by the D.O. Mills Expedition, Period 1903–1906," *Publications of the Astronomical Society of the Pacific* 23, no. 139 (December 1911); W.W Payne, "The Mills Expedition from Lick Observatory," *Journal of Popular Astronomy* 15 (1907).

[24] D.O. Mills Expedition, Correspondence—W.W. Campbell to D.O Mills, September 27, 1907 UA 36. Ser 04. Box 8 Folder 3, *Lick Observatory Records*, Special Collections, Mary Lea Shane Archives, University of California Santa Cruz.

[25] D.O. Mills Expedition, Correspondence—W.W. Campbell to D.O Mills, September 27, 1907 UA 36. Ser 04. Box 8 Folder 3, *Lick Observatory Records*, Special Collections, Mary Lea Shane Archives, University of California Santa Cruz.

this purpose, they tried to diversify their publications. Campbell was particularly proud of one of them: "Samples of results for three individual stars, and for Venus, Mars, and the Moon, are given on page 61 in order to show the great accuracy obtained in the work of the Expedition. The chart on page 62 shows at a glance the distribution of the stars which the Expedition studied."[26] This was what Campbell reported to Mills in 1907, but to that we have to add other publications in collaboration with other scientists, in which he elaborated a stellar catalogue and analysis of the Comet A.[27] In fact, a few years later they would publish a general catalogue of stars, which included those unknown stars of the south that the Lick Expedition was delivering to the scientific world.[28]

Campbell also had to ensure a perfect synchronicity between the southern station and Mount Hamilton. He also needed to secure equal accuracy. The perfection in that assemblage had much to do with the equipment. As Wright explained, the combination of the reflecting telescope of 94 cm, the Cassegrain focus and the three-prism spectrograph secured trustworthy results. In 1904 they already had 308 accurate spectrograms, comparable to those obtained at Mount Hamilton.[29]

Little by little, questions related to the sidereal problem began to seem resolvable. At least astronomers felt they were getting closer to understanding the structure of the galaxy. They needed every piece of information they had on the familiar northern skies as well as the new and mysterious southern universe. There, looking up from the southern world, they began confirming theories on their bright companions in the dark nights of observation.

[26]D.O. Mills Expedition, Correspondence—W.W. Campbell to D.O Mills, September 27, 1907 UA 36. Ser 04. Box 8 Folder 3, *Lick Observatory Records*, Special Collections, Mary Lea Shane Archives, University of California Santa Cruz.

[27]William W. Campbell, R.G. Aitken, and J.D. Maddrill, "The Observatory of the D.O. Mills Expedition to Chile," *Publications of the Astronomical Society of the Pacific* 16, no. 96 (June 1904): 144–145.

[28]William H. Wright, "Radial Velocities of 150 Stars South of Declination—20°. Determined by the D.O. Mills Expedition Period 1903–1906 & General Catalogue"; *D.O. Mills Expedition. Cerro San Cristóbal, Santiago, Chile. From Its Organization in December 1900 to March 1906*, ed. W.W. Campbell and W.H. Wright, Vol. IX (Sacramento: W. W. Shannon Superintendent State Printing, 1907), 328–343.

[29]William H. Wright, "On Some Results Obtained by the D.O. Mills Expedition to the Southern Hemisphere," *Lick Observatory Bulletin* 60 (1904).

> Three other articles relate to investigations that he has made on the orbit of stars shown by the expedition to be, not ordinary single stars, but stars shown by massive invisible companions which causes the visible stars to revolve in orbits. The Expedition has shown for the southern stars what the Mills Spectrograph on Mount Hamilton first showed for northern stars, that at least one star in every six or seven is attended by a massive invisible companion comparable in size with the bright star itself, the bright star and invisible companion in each case necessarily revolving around each other.[30]

The Lick astronomers later added nebulae to their research.[31] These discoveries, technological innovations, and widening of the knowledge that humans had of the galaxy had other consequences. It was essential to inform Mills of all of this to strengthen and perhaps increase the funding of the expedition. Campbell's self-promotion before his benefactor was anything but coy: "You may have noticed in the papers of February 10th that the Gold medal of the Royal Astronomical Society of London was conferred upon me, and that Ambassador Reid attended the meeting of the Society I received the medal for transmission through the State Department to me."[32] This was not trivial information. Besides this acknowledgment, Campbell mentioned the ambassador who would later marry Mills's daughter, Elizabeth. Mills might have wanted to know about such distinctions.

> The address of the President of the Society has not yet reached me; but he very appropriately mentioned your name on that occasion, because every one of the observations for which the Medal was bestowed was made either with the Mills Spectrographs on Mount Hamilton, attached to the 36-inch refracting telescope, or with the instruments of the Mills Expedition to Chile.[33]

[30] D.O. Mills Expedition, Correspondence—W.W. Campbell to D.O Mills, September 27, 1907 UA 36. Ser 04. Box 8 Folder 3, *Lick Observatory Records*, Special Collections, Mary Lea Shane Archives, University of California Santa Cruz.

[31] "Studies of the Nebulae," *Publications of the Lick Observatory*," Vol. XIII (Berkeley: University of California Press, 1918).

[32] D.O. Mills Expedition, Correspondence—W.W. Campbell to D.O Mills, February 26, 1906, UA 36. Ser 04. Box 8 Folder 3, *Lick Observatory Records*, Special Collections, Mary Lea Shane Archives, University of California Santa Cruz.

[33] D.O. Mills Expedition, Correspondence—W.W. Campbell to D.O Mills, February 26, 1906, UA 36. Ser 04. Box 8 Folder 3, *Lick Observatory Records*, Special Collections, Mary Lea Shane Archives, University of California Santa Cruz.

Other references to Mills's largesse were of course replicated in reports on the astronomers' work in the southern station. As a colleague would say, "Director Campbell is very fortunate in having behind his planning and its execution so able, intelligent and generous a man as D.O. Mills. It will certainly be of interest to all our readers to know what this kind of benevolence means to a man when he undertakes to carry through an enterprise of this kind."[34] As a leading figure, Mills would be embedded in the acknowledgments of astronomy in the early twentieth century. Campbell did not hesitate to make his gratitude explicit. "I am exceedingly glad that the President did mention your name in this way; for it is known to astronomers everywhere that my success in this work was made possible by your help."[35]

The next year, in 1907, Campbell added news regarding other acknowledgments on the work done that combined data from California and Chile, this time from the United States. He received a medal from the US National Academy of Sciences for his work with the Mills Spectrographs.[36] By coincidence, Campbell had forgotten to tell Mr. Mills about that important award that granted him national prestige. The ceremony had taken place barely three hours before the San Francisco earthquake, in April 1906. Campbell was in Washington, DC but changed plans and returned immediately to California without meeting Mills on the East Coast.

To have more standing within the United States was indeed important. The opportunity arose again for Lick's director to express both publicly and privately his gratitude to the patron of his scientific development. Thus, there were other moments in which Campbell communicated their scientific results, among which he underlined his conference at Yale University.[37]

[34] Payne, "The Mills Expedition," 523.

[35] D.O. Mills Expedition, Correspondence—W.W. Campbell to D.O Mills, February 26, 1906, UA 36. Ser 04. Box 8 Folder 3, *Lick Observatory Records*, Special Collections, Mary Lea Shane Archives, University of California Santa Cruz.

[36] D.O. Mills Expedition, Correspondence—W.W. Campbell to D.O Mills, January 31, 1907, UA 36. Ser 04. Box 8 Folder 3, *Lick Observatory Records*, Special Collections, Mary Lea Shane Archives, University of California Santa Cruz.

[37] William W. Campbell, *Stellar Motions, with Special Reference to Motions Determined by Means of the Spectrograph* (New Haven: Yale University Press, 1913). This publication was part of Mrs. Hepsa Ely Silliman Memorial Lectures.

The combined work of Chile and California seemed to be fulfilling expectations. The southern skies revealed the enormous knowledge yet to be discovered. Chile continued to deliver surprises and not only because of the social adaptation of the astronomers. The idea of making the southern observatory a permanent station continually floated in the scientists' thoughts. If that was going to be a plan to explore, then they needed to know what would be possible for observation beyond Santiago. They knew there was a desert on the northern area of the country. Locals called it the "Deserted Atacama."

4.4 Dr. Curtis's Adventure in Atacama

A few years after the astronomers' arrival in Chile, plans to establish a permanent location in the southern hemisphere gained strength. If such plans had been in Campbell's mind since the beginning, he did not discuss them in his negotiations with the University of California over financing and permission. Yet, once Mills had consented to prolong his financing of the expedition, and once Campbell had the materials to justify the profitability of the project, a permanent station became the subject of serious consideration.

Curtis knew in detail all the advances and problems that the various initiatives in the south had encountered so far and discussed them in an article published in 1909. Up to that date, the largest project to map the universe, the *Carte du Ciel*, was already showing its shortcomings. This project was about astronomy of position and, for Curtis, the program's scope was so vast that even in the northern hemisphere this project had not made the progress anticipated at the time of its inception.[38] Hence, initiatives in the southern hemisphere became more important as they were determined to solve astronomical questions concerning those unknown skies visible from the south. The Mills Expedition was one of them. The work in Chile might be the key to such vast questions as the movement of the solar system throughout the universe and the structure of the Milky Way.

Since their arrival in Chile, the astronomers had explored a few sites, but near Santiago. When arriving, Dr. Palmer went to see San Felipe—about 100 kilometers north from Santiago-, and regretted that there was

[38] Heber D. Curtis, "Astronomical Problems of the Southern Hemisphere," *Publications of the Astronomical Society of the Pacific* 21, no. 129 (December 1909): 231–244.

no time to go to Copiapo.[39] During those first years, Wright had been thinking of a possible transfer of the observatory, due perhaps to his conversations with Campbell. For his part, the Lick Director had been evaluating the possibility of a new site with Mills; as early as 1905 he wrote, "Mr. Wright thinks that a location few hundred miles north of Santiago can be found which combines a larger proportion of clear nights than Santiago has, with the other favorable and necessary conditions possessed by the present location."[40]

The Chilean press, expressing premature and excessive confidence, was already supposing that the gringos were going to stay permanently. "Before choosing the site for the southern hemisphere station, Dr. Curtis will proceed to investigate the atmospheric conditions of one or two points of Chile and perhaps of some of the region of Córdoba in the Argentine Republic."[41] In fact, at the end of his assignment as acting astronomer in charge of the Mills Expedition, Curtis did not evaluate sites in Argentina, but around Copiapo in northern Chile.

The astronomers had already been discussing what to do with the equipment at the San Cristóbal if they decided to transfer. Campbell wrote to Mills:

> The educational officials of the Chilean Government have expressed their desire to purchase the steel dome and building, in case we move to another station. The Chilean Naval *Charge d'Affairs* at Washington came to Mount Hamilton last summer to see about buying the entire equipment now on Mount San Cristóbal. It would be advantageous to sell the dome and buildings rather than to dissect and move them; but I hope we shall not want to sell the telescope and spectroscope.[42]

[39] Beverly Hard, "Notes on the Directors of the D.O Mills Expedition to Chile," 1972. Expeditions, UA 36. Ser 04. Box 12 Folder 6, *Lick Observatory Records*, Special Collections, Mary Lea Shane Archives, University of California Santa Cruz, 7.

[40] D.O. Mills Expedition, Correspondence—W.W. Campbell to D.O Mills, January 21, 1905, UA 36. Ser 04. Box 8 Folder 3, *Lick Observatory Records*, Special Collections, Mary Lea Shane Archives, University of California Santa Cruz.

[41] "El observatorio del San Cristóbal."

[42] D.O. Mills Expedition, Correspondence—W.W. Campbell to D.O Mills, January 21, 1905, UA 36. Ser 04. Box 8 Folder 3, *Lick Observatory Records*, Special Collections, Mary Lea Shane Archives, University of California Santa Cruz.

Chilean authorities saw what the Mills Expedition could offer to the country. But years had to pass before any purchase of the Mills Observatory could be considered.

Dr. Curtis led the exploration of the north of Chile, about which the Americans knew very little. In April of 1909, he went to Valparaíso and got on a steamship to Caldera. The distance was so vast that a road trip would have been unthinkable. In addition to consuming precious time and energy, the voyage would have imperiled the perfect state of the scientific instruments needed to assess the site. Curtis wrote a detailed and complete report of everything he observed and found. The resulting *Curtis Report* was, perhaps, the first to describe the possibilities and conditions for astronomical observation of the Atacama Desert. Without Curtis even suspecting it, decades later, in a vastly different world, great international astronomical corporations would also explore the desert to find the ideal place for their astronomical research. The area eventually will become home for approximately 70% of the global observation surface.[43]

Curtis disembarked at Caldera, some 900 kilometers north of the capital of Chile, in April of 1909. The *Curtis Report* on the conditions of the Atacama begins with a detailed description of the area's infrastructure, decisive for the success of any astronomical expedition. Curtis described life in that remote and deserted place. If the plan to settle in the area paid off, all these details would be important. On Caldera, Curtis wrote, "There are several places where lodging and meals can be secured; the best of these, the Hotel Suiza, is very bad. At Caldera are located extensive copper smelters and repair shops of the Copiapó Railroad; these are well equipped and any castings, repairs or machine work could be executed there."[44] Oddly enough, Curtis also noted other features of the area such as "two fair beaches for bathing" and "numerous little bath houses which are rented to families spending a part of the summer there for beach and surf bathing."[45] Apparently vacation

[43] Michele Catanzaro, "Big Players / Chile: Upward Trajectory," *Nature* 510 (June 12, 2014): 204–205.

[44] Heber D. Curtis, "Report on Astronomical Conditions in the Region About Copiapo," D.O. Mills Expedition: Report on Site Survey Near Copiapo. April 17, 1909, UA 36. Ser. 04. Box 8 Folder 12, *Lick Observatory Records*, Special Collections, Mary Lea Shane Archives, University of California Santa Cruz, 1.

[45] Curtis, "Report on Astronomical Conditions," 1.

practices had arrived in Chile by then and locals were adopting summer leisure time at the beach.

Curtis made his way inland, about 80 kilometers away, toward the city of Copiapo. The connection was efficient, as the train departed every day at 1:30 p.m. Curtis found out this was the first railroad constructed in South America. He also learned it was a private road, owned by English capital, and "rarely if ever grants favors or passes even to managers of mines who ship much freight; it has a bad reputation in this respect and it is more than doubtful whether any favors, passes or reductions in tariffs could be secured for a scientific expedition."[46]

The train connection to the interior area was important. Without it there could be no exploration, much less an astronomical mission. Yet another train went up to the mountains three times a week from Copiapo, the only large town in the area, by then, with some ten thousand inhabitants. "All the other places are generally only small stopping places at various haciendas. Tierra Amarilla and Puquios are the largest, and a doubtful four hundred is claimed as the population of Puquios."[47] In fact, Curtis was heading to Puquios. But he noted other features during his short stay in Copiapo, the "big" town in the area. For Curtis, Copiapo was boring and its claim as a "city" was dubious. Even so, he was impressed to find hotels, schools for boys and girls, and even technical schools. Curtis detailed other interesting characteristics, such as the presence of a small English speaking community. Despite the size of the town and the rustic environment, Curtis was surprised about expenses: "I gather that the cost of living in Copiapo is about the same as in Santiago, i.e., pretty high."[48]

Curtis was clearly thinking of the possibilities of family life of astronomers that would need to draw on these facilities. Probably that is why he also stated his surprise at the cost of living. For the same reason, he inquired about the availability and cost of servants and gathered information about basic services, such as water. In the desert, there was no system of water provision and that was of crucial relevance for quality of

[46] Curtis, "Report on Astronomical Conditions," 1.

[47] Curtis, "Report on Astronomical Conditions," 2–3.

[48] Curtis, "Report on Astronomical Conditions," 3.

life. For astronomical purposes, he also reported that there was no electric light there.[49]

After these initial impressions, Curtis informed about the weather and observation conditions. Fog, which gathered in the valley during the night and remained there until the following morning, was a major problem. He talked to people whose occupations made them get up very early in the morning and found out that every one of them mentioned these thick morning clouds as an everyday feature. This information was important and indicated that fog was a daily characteristic of the Copiapo region. "These clouds are driven from the sea, sometimes low enough to be felt as a fog, at other times higher, but almost always thick enough to cover all the peaks around."[50] This was a deal-breaker that precluded even thinking of setting an observatory there.

Curtis found out that the local high school, the Liceo de Copiapo, kept consistent meteorological records that confirmed that 70% of the time Copiapo and its surroundings were covered by thick fog at dawn.[51] With this regular night fog coming in from the sea, these impressions brought to mind San Jose, California, downhill from Mount Hamilton. "Frequently this thick fog is accompanied by a fine mist or drizzle, especially in winter." But, there was a big difference since they were in the driest desert on earth. "Rain will average once or twice a year; there has been <u>none</u> for the past two years."[52]

The familiarity that Wright had felt upon disembarking in Valparaíso years ago arose again. Curtis, now in the north of Chile, sought his references in his beloved California to explain the landscapes of this reflection. But he had to underline the absolute absence of rain, and therefore, the scarcity of water. He began to understand how dry a place could be and how that contrasted with the impenetrable fog. The only solution to escape the fog seemed to be to go up the mountains. But fog was not the only concern; there was also the daily range in temperature, which, according to a teacher at the Liceo, averaged close to 15 °C and more during winter.[53]

[49] Curtis, "Report on Astronomical Conditions," 4

[50] Curtis, "Report on Astronomical Conditions," 4.

[51] Curtis, "Report on Astronomical Conditions," 5–7.

[52] Curtis, "Report on Astronomical Conditions," 8. Underlined in the original.

[53] Curtis, "Report on Astronomical Conditions," 8.

All these factors pushed Curtis to explore other areas. "Very laudatory reports were given me of the great clearness at Puquios, the northern terminus, 143 kilometers from Caldera. I went up there on April 8th; the trip takes about four hours and is up through a barren valley-cannon all the way. I remained there or in the vicinity for three days, till the next train back."[54] It was a small town on a sandy plain. Basic services were "primitive," as Curtis wrote, although it functioned as a distribution center for supplies for copper mines in the mountains. Curtis again used references to the United States that helped him communicate his impressions to his boss of this rustic, strange place. "Puquios looks much like some Arizona adobe town stripped of its sagebrush surroundings and with higher and more rugged mountains."[55] The Chilean desert, however, was much deeper and more hermetic.

It was therefore surprising to find in the little town of Puquios a butchery and even a pharmacy. The water these stores used was "piped down from the Llano de Varas." According to Curtis, the water's quality was not very good, but still locals used it. The town was some sort of distribution center: "The main life in the town comes from the trains of mule carts bringing in copper ore and carrying back coke and supplies, or train of burros going and coming from the most inaccessible spots by narrow zigzag trails."[56] In the midst of such scarcity, it was remarkable that there were small human settlements attached to the copper mines. It was even more surprising that there were telephone lines between Copiapo and Caldera. Mines such as the Copiapo Mining Co., known as the *Compañía Inglesa de Minas*, were in British hands; Americans owned others. This was important not only because of the possibility of finding English speaking communities, but also because there was some infrastructure, such as railroads and projects for constructing water supply systems, and expert knowledge of mining engineers working in the area. "It is true that such a site would be quite isolated, but the copper companies seem to have had no difficulties in bringing in their heavy machinery, and supporting eight or ten Englishmen and eighty to one hundred and fifty laborers from Copiapo as a base."[57]

[54] Curtis, "Report on Astronomical Conditions," 10.

[55] Curtis, "Report on Astronomical Conditions," 10.

[56] Curtis, "Report on Astronomical Conditions," 10.

[57] Curtis, "Report on Astronomical Conditions," 14.

Curtis reported on all aspects—from living conditions to technical feasibility to the possibility of settling in Llano de Varas. The descriptions included proximity to cargo terminals, the number of railroads, whether other roads were in the area, construction difficulties, water supply, and English speaking neighbors. Just as he had done before in Copiapo, he added availability of education. "There is a little native school in the canyon at La Descubridora and another at Puquios, but not suitable for 'white men'."[58] The school for local children was, in his eyes, not suitable for American kids. These details suggest that Curtis was wondering if family life was possible in Puquios, or if only single men could go to the desert. He tried to minimize the loneliness of the Chilean desert. He thought the Llano would doubtless be much better than the nitrate desert, farther north and much more remote.

Northern Chile also featured the magnificent Cordillera de los Andes, which could offer an escape from the morning fog. He learned Copiapo was the last fertile spot for a thousand miles north, except for some canyons of the Andes which had morning fog. Curtis was surprised by the potential of the Copiapo mountains: "The sky at the Llano impressed me as very blue and pure; no whitening toward the horizon."[59] Other options in the same zone had the major problem of the fog. "All along this coast, just as in California, an extremely heavy fog (chamacha?) rolls in and fills up everything to an altitude of about 4000 ft."[60] He may have been not able to reproduce the local word for the fog, *camanchaca*. But, he did point out how good the conditions would be in the "nitrate desert," even farther north than the Copiapo area. "Nearly any point on the nitrate desert at an altitude above 4000 feet would doubtless have the same general climate as Puquios,"[61] but that altitude implied several other problems back then. Decades before the current cutting-edge observatories of today were erected there, Curtis understood the area's potential for astronomy.

[58] Curtis, "Report on Astronomical Conditions," 17.

[59] Curtis, "Report on Astronomical Conditions," 20.

[60] D.O. Mills Expedition, Correspondence—Herber D. Curtis to W.W. Campbell, "Trip to Puquíos". June 29, 1916. UA 36. Ser. 04. Box 7 Folder 9, *Lick Observatory Records*, Special Collections, Mary Lea Shane Archives, University of California Santa Cruz.

[61] D.O. Mills Expedition, Correspondence—Herber D. Curtis to W.W. Campbell, "Trip to Puquíos". June 29, 1916. UA 36. Ser. 04. Box 7 Folder 9, *Lick Observatory Records*, Special Collections, Mary Lea Shane Archives, University of California Santa Cruz.

The area seemed apt for astronomical purposes, even compared to South Africa. Curtis mentioned that an Englishman in charge of a mine in the local area, Mr. Stark, had "lived here two years and lived several years near Johannesburg in South Africa. His unhesitating opinion was that Puquios was very much clearer than Johannesburg."[62] Curtis recorded the impressively clear skies in Puquios. On average, there were only three cloudy days in eight months. "Not a cloud was visible day or night during my stay here and at Puquios (…) the impression of all with whom I talked is that the overwhelming majority of all days and nights are perfectly clear."[63] Considering all the information he gathered, Curtis was confident enough to predict that a site in the Llano de Varas would average 300 clear nights per year.[64]

Still, it would be an extremely unusual place for astronomical observation. His words convey the solitude, the desolation, and the stark conditions of the Atacama Desert. One of Curtis' most pressing concerns was the area's isolation, despite the presence of English speaking people, which for him was definitely a positive factor. He also gathered some information of the place in between Santiago and Copiapo, called Coquimbo, which he described as "a dull little place from all accounts. Region around is fertile. I did not visit, but found out much from a gentleman who has lived there and at Ovalle for several years; climate more like that of Santiago; sea fog reaches here. Railroad is property of the government."[65]

Despite the impressive clarity of Atacama, his "down to earth" judgment led Curtis to conclude that the clear nights in Santiago were more than enough. Indeed, this—along with the lack of funding for another observatory in an area even more difficult to reach—was the reason for staying on San Cristóbal. There was an additional difficulty for astronomical observation. Although the thick morning fog and frequent mists in Copiapo did not get in the higher elevation of Puquios, the valley was too narrow for an astronomical station.[66]

After his thorough exploration of the Copiapo area, Curtis was convinced that Santiago was the best place for the observatory.

[62] Curtis, "Report on Astronomical Conditions," 11.

[63] Curtis, "Report on Astronomical Conditions," 13.

[64] Curtis, "Report on Astronomical Conditions," 14.

[65] Curtis, "Report on Astronomical Conditions," 19.

[66] Curtis, "Report on Astronomical Conditions," 11.

Although, his overall evaluation was positive, he suspected that Campbell would not want to change the current location. "To 'live' at the same time with astronomical work, there is nothing in South America like Santiago; site near Copiapo would be a very distant second." On this point, Curtis was categorical: "If a site must be near a center of population I doubt if anything in the southern hemisphere would be better than the present location on San Cristóbal."[67] He was emphatic about the scarce possibilities for making astronomical observation in the Coquimbo area (the southern, semiarid area of the desert).

Nonetheless, his report tells about life in the desert beyond the nitrate camps in the early twentieth century. The information was valuable for the time, and continues to illustrate what the region was like more than a hundred years ago. Curtis was the first to evaluate, scientifically and accurately, the possibility of making astronomical observations in the Atacama Desert.

Decades later, with the growth of astronomy in Chile during the 1960s, Curtis report became the subject of renewed attention. In the late 1990s, Duerbeck, Osterbrock, Barrera, and Leiva said, "The Lick experience was important when planning began for the large European and American southern hemisphere observatories in the 1950s."[68] Other authors have reiterated the sentiment as well.[69] According to them, the Curtis report was crucial to the installation of international astronomical holdings—both Americans and Europeans—in Chile in the early 1960s. Other records also underscore this link, indicating some unspecified relation between the Curtis Report and the decision of the Association of Universities for Research in Astronomy (AURA) to settle in Chile, in the Coquimbo area.[70]

[67] Curtis, "Report on Astronomical Conditions," 18.

[68] Hilmar Duerbeck, et al. "Halfway form La Silla to Paranal – in 1909," *The Messenger* 95 (March 1999).

[69] Ricardo Leiva, "Atacama: a 100 años del informe Curtis," *Revista de Humanidades* 3 (2010): 21.

[70] "But six years later, at the end of Dr. Curtis' term he discussed the possibilities of Copiapo, of Antofagasta and Iquique and even of Tacna, which is southern Peru. It is interesting that the port of Coquimbo, less than three hundred miles north, is also a place he mentioned, because it is very near La Serena, which is the headquarters for the Cerro Tololo Inter American Observatory which has been established by the Association of Universities for Research in Astronomy and supported by the National Sciences Foundation," Hard, "Notes on the Directors," 7.

Despite there is no clear proof of this relation (at least for now), we can state Curtis helped situate the Atacama Desert as a place with great potential for astronomical observation. In a world that had significantly changed, the obstacles Curtis saw in the early twentieth century were no longer a problem by 1960. It is hard to say what influence the report had on the subsequent development of astronomy in Chile. But clearly the semiarid area of Coquimbo and the desert that Curtis visited and assessed fostered the beginnings of transnational astronomy in Chile in the middle of the Cold War.[71]

Curtis continued his work at the Lick Observatory until 1920, when he left for Pittsburgh to be the director of the Allegheny Observatory. Afterwards he was appointed director of the University of Michigan Observatories. While moving from Lick to Allegheny, he participated in the famous Shapley-Curtis debate at the Smithsonian Museum of Natural History of 1920. The two astronomers, Curtis and Harlow Shapley, debated the scale of the universe. Curtis insisted that separate galaxies such as Andromeda existed.[72] Years later, evidence would prove Shapley right about the size of our galaxy and Curtis right about the number of galaxies that make up the universe.

Bibliography

Campbell, William W. "Will Determine Whether Solar System Speeds in Its Flight Through Space," D.O. Mills Expedition, February 23, 1903, UA 36. Ser 04. Box 8 Folder 1, *Lick Observatory Records*, Special Collections, Mary Lea Shane Archives, University of California Santa Cruz.

Campbell, William W., R.G. Aitken, and J.D. Maddrill, "The Observatory of the D.O. Mills Expedition to Chile," *Publications of the Astronomical Society of the Pacific* 16, no. 96 (June 1904): 144–145.

Campbell, William W. "Organization and History of the D.O. Mills Expedition to the Southern Hemisphere," *D.O. Mills Expedition. Cerro San Cristóbal, Santiago, Chile. From Its Organization in December 1900 to March 1906*,

[71] Bárbara Silva, PI, *Stars and Galaxies from the South. Chile and Its Scientific—Astronomical Insertion in the Global Cold War, 1962–1973*, FONDECYT Research Project 3170099.

[72] Virginia Trimble, "The 1920 Shapley–Curtis Discussion: Background, Issues, and Aftermath," *Publications of the Astronomical Society of the Pacific* 107 (December 1995): 1133–1144.

ed. W.W. Campbell and W.H. Wright, Vol. IX, 5–14. Sacramento: W. W. Shannon Superintendent State Printing, 1907.

Campbell, William W. "Notes on the D.O. Mills Expedition to the Southern Hemisphere," *Publications of the Astronomical Society of the Pacific* 23, no. 135 (February 1911): 52–54.

Campbell, William W. *Stellar Motions, with Special Reference to Motions Determined by Means of the Spectrograph.* New Haven: Yale University Press, 1913.

Catanzaro, Michele. "Big Players / Chile: Upward Trajectory," *Nature* 510 (June 2014): 204–205.

Curtis, Heber D. "Recent Changes at the Observatory of the D.O. Mills Expedition," *Publications of the Astronomical Society of the Pacific* 19, no. 116 (October 1907): 227–233.

Curtis, Heber D. "Temperature Control for Silvered SPECULA," *Astrophysical Journal* 26 (November 1907): 256–262.

Curtis, Heber D. "Report on Astronomical Conditions in the Region About Copiapo," D.O. Mills Expedition: Report on Site Survey Near Copiapo. April 17, 1909, UA 36. Ser. 04, Box 8 Folder 12, *Lick Observatory Records*, Special Collections, Mary Lea Shane Archives, University of California Santa Cruz.

Curtis, Hebert D. "Recent Progress in the Work of the D.O. Mills Expedition to the Southern Hemisphere," *Publications of the Astronomical Society of the Pacific* 21, no. 128 (October 1909): 199–201.

Curtis, Heber D. "Astronomical Problems of the Southern Hemisphere," *Publications of the Astronomical Society of the Pacific* 21, no 129 (December 1909): 231–244.

D.O. Mills Expedition, UA 36. Ser 04. Boxes 1 to 12, *Lick Observatory Records*, Special Collections, Mary Lea Shane Archives, University of California Santa Cruz.

Duerbeck, Hilmar W., D.E. Osterbrock, L. Barrera, and R. Leiva. "Halfway form La Silla to Paranal – in 1909," *The Messenger* 95 (March 1999): 34–37.

"El Observatorio del San Cristóbal," *Revista Zig-Zag*, May 13, 1906.

Frost, Edwin B. "Radial Velocities of 150 Stars South of Declination—20° Determined by the D.O. Mills Expedition, Period 1903–1906," *Publications of the Astronomical Society of the Pacific* 23, no. 139 (December 1911): 257–260.

Hard, Beverly. "Notes on the Directors of the D.O Mills Expedition to Chile," 1972. Expeditions, UA 36. Ser 04. Box 12, Folder 6, *Lick Observatory Records*, Special Collections, Mary Lea Shane Archives, University of California Santa Cruz.

Leiva, Ricardo. "Atacama: a 100 años del informe Curtis," *Revista de Humanidades* 3 (2010): 10–22.

New York Times, March 24, 1905.

Payne, W.W "The Mills Expedition from Lick Observatory," *Journal of Popular Astronomy* 15 (1907): 522–527.

"Studies of the Nebulae," *Publications of the Lick Observatory*, Vol. XIII (Berkeley: University of California Press, 1918).

Trimble, Virginia "The 1920 Shapley–Curtis Discussion: Background, Issues, and Aftermath," *Publications of the Astronomical Society of the Pacific* 107 (December 1995): 1133–1144.

Wilson, E. "Roscoe Frank Sanford, 1883–1958," *Publications of the Astronomical Society of the Pacific* 70, no. 415 (1958): 360–363.

Wright, William H. "On Some Results Obtained by the D.O. Mills Expedition to the Southern Hemisphere," *Lick Observatory Bulletin* 60 (1904): 3–5.

Wright, William H. "Radial Velocities of 150 Stars South of Declination—20°. Determined by the D.O. Mills Expedition Period 1903–1906 & General Catalogue," *D.O. Mills Expedition. Cerro San Cristóbal, Santiago, Chile. From Its Organization in December 1900 to March 1906*, ed. W.W. Campbell and W.H. Wright, Vol. IX, 71–347. Sacramento: W. W. Shannon Superintendent State Printing, 1907.

CHAPTER 5

Changing Years for the Mills Expedition

Abstract This chapter studies how the mission changed as years went by, especially after they had to look for new sources of funding. It also gives a new perspective on what the Great War meant for science, and how Chilean society was changing, mainly through its urban history—that also affected the observatory, since the place they chose was transformed into a public park. This chapter also describes plans for finishing the mission and the departure of the Lick's team. The Americans decided to sell the Observatory to Chile's Catholic University, and then again cultural differences challenged the negotiations. The astronomers left the country with great sorrow, after living and working in that far away country.

Keywords Great War · Urban history · Catholic University of Chile

The tasks of the California astronomers in Santiago moved forward and continued to yield results. Such progress was fundamental in convincing Mills to extend the expedition's funding. Several American astronomers passed through the mission in Chile. Despite the difficulties, everything seemed to work according to plan. Although a cultural gap remained, the gringos adapted to life in Santiago society and on San Cristóbal hill. Still, the mission was about to change, on many different levels. Individual lives, the quest for funds, international relations, and even

B. K. Silva, *Astronomy at the Turn of the Twentieth Century in Chile and the United States*, Palgrave Studies in the History of Science and Technology, https://doi.org/10.1007/978-3-030-17712-6_5

urban development in Santiago were all decisive for the future of the Mills expedition in the 1910s. As discussed, Campbell must have had the intention of making the Mills Expedition into a permanent station from the beginning, or at least, not long after the mission settled in Chile. By 1905 Campbell was openly discussing this possibility with his benefactor.

> My strong desire is to see the Expedition converted into a permanent observatory, to take up one great problem after another along the lines for which it is equipped. If the leading astronomers were asked for their views on this subject, I feel sure everyone would urge very strongly that this be done, and that in no other way could a given sum promote the science of astronomy more efficiently. Should you like to have their views, I would gladly undertake to obtain them.[1]

In the years to come, this subject would appear once and again in their correspondence. Mills continued funding the expedition, but it was not decided it would become a permanent station. Years passed, and there were some unexpected news Campbell was about to receive that would jeopardize the expedition's future.

5.1 The Donor's Death

Campbell was determined but he needed Mr. Mills's continued sponsorship. One strategy he used to convince Mills was to create an association with the University of California. To this end, Campbell brought on Benjamin Wheeler, the president of the university. With the idea of a permanent station in mind, Campbell wrote to Mills. "I very respectfully ask you to endow a Mills Observatory in the Southern Hemisphere. With the approval of the president Wheeler, I would suggest that the proposed observatory should form a part of the Astronomical Department of the University of California, just as the Lick Observatory is in that Department."[2] Campbell's plan was to make both observatories equally important and autonomous, such that the decision-making would

[1] D.O. Mills Expedition, Correspondence—W.W. Campbell to D.O Mills, January 21, 1905, UA 36. Ser 04. Box 8 Folder 3, *Lick Observatory Records*, Special Collections, Mary Lea Shane Archives, University of California Santa Cruz.

[2] D.O. Mills Expedition, Correspondence—W.W. Campbell to D.O Mills, January 21, 1905, UA 36. Ser 04. Box 8 Folder 3, *Lick Observatory Records*, Special Collections, Mary Lea Shane Archives, University of California Santa Cruz.

not only come from the Lick. "The two institutions may cooperate in extending their observations to the entire celestial sphere, on the same homogenous plan."[3] Mills's patronage of the southern observatory would thus be much more than a minor donation to a subsidiary branch. It would be a significant contribution to a major astronomical center, with all the prestige of the university.

Moreover, the plan was consistent with Campbell's theory that the sidereal problem required observations of the entire sky and therefore coordination between both observatories: "Permanent observations are woefully lacking in the southern hemisphere. They are unable to do their share on the southern third of the sky."[4] In addition, by carrying through this plan, he would contribute in consolidating the presence of California in southern astronomy. Once again, he quoted the Carnegie Yearbook to strengthen his argument, where he, along with astronomers Boss and Hale, summarized expert opinion when consulting "to the leading astronomers of the world." He added

> It does not seem at all surprising that our colleagues, with emphatic unanimity, should have expressed the opinion that additional astronomical effort at the present time can be expended to the best advantage in the southern hemisphere. The British Government, in maintaining an astronomical observatory of high rank at the Cape of Good Hope, can scarcely be criticized for not doing more in that quarter of the world.[5]

As the report noted, there was a permanent observatory in South Africa, a location Curtis compared to Atacama. Due to their colonial system, the position of the United Kingdom was obviously a great privilege for installing a scientific mission there. Campbell argued that, of all the projects that could be carried out in the south, spectroscopic

[3] D.O. Mills Expedition, Correspondence—W.W. Campbell to D.O Mills, January 21, 1905, UA 36. Ser 04. Box 8 Folder 3, *Lick Observatory Records*, Special Collections, Mary Lea Shane Archives, University of California Santa Cruz.

[4] D.O. Mills Expedition, Correspondence—W.W. Campbell to D.O Mills, April 4, 1905, UA 36. Ser 04. Box 8 Folder 3, *Lick Observatory Records*, Special Collections, Mary Lea Shane Archives, University of California Santa Cruz.

[5] D.O. Mills Expedition, Correspondence—W.W. Campbell to D.O Mills, April 4, 1905, *Carnegie Institution Year Book*, no. 3, p. 25 UA 36. Ser 04. Box 8 Folder 3, *Lick Observatory Records*, Special Collections, Mary Lea Shane Archives, University of California Santa Cruz.

observations were most important. He also reported they were already building a spectroscope to observe fainter stars.

After several negotiations and initial reluctance from Mills, Campbell managed to convince his patron to provide five additional years of financing for the expedition, which would allow the project to operate until 1911. But Campbell began to lobby for more time in late 1909. Campbell wrote to Mills details of the expedition's fantastic progress. "The results of the expedition's work I regard as of the highest astronomical value. Their publication is anxiously awaited by astronomers in all countries."[6] He never got an answer. Mills had died in January of 1910, leaving the Lick astronomers in deep uncertainty.[7]

Campbell began to speak with the heirs, particularly with Mills's son Ogden, about continuing the work in the south. These discussions were not easy. Campbell did not know how to gauge Ogden's commitment to scientific research. The son never had much of a relationship with the Lick Observatory, beyond his father's connections. Still, it seemed likely that, in keeping with American customs of the time, the heir would honor his father's commitments. Campbell sent Ogden detailed notes on the agreements established with the magnate regarding how much money had been transferred and by what means.[8]

The heir, who was now administering his father's businesses and vast fortune, was not content to merely follow Campbell's instructions. Although Ogden sent some money for the expedition, he wanted to have "a little clearer understanding about the arrangement" before proceeding and therefore requested a meeting with Campbell.[9] The agreement was not going to continue without Campbell describing to the younger Mills the previous seven years of work. Explaining fundamentals, such as the meaning of and potential for spectography, in layman

[6] D.O. Mills Expedition, Correspondence—W.W. Campbell to D.O. Mills, December 23, 1909 UA 36. Ser 04. Box 8 Folder 3, *Lick Observatory Records*, Special Collections, Mary Lea Shane Archives, University of California Santa Cruz.

[7] William W. Campbell, *A Brief Account of the Lick Observatory of the University of California* (Berkeley: University of California Press, 1931), 21.

[8] D.O. Mills Expedition, Correspondence 1910–1923—W.W. Campbell to O. Mills, January 18, 1910. UA 36. Ser 04. Box 8 Folder 1, *Lick Observatory Records*, Special Collections, Mary Lea Shane Archives, University of California Santa Cruz.

[9] D.O. Mills Expedition, Correspondence 1910–1923—O. Mills to W.W. Cambpell, January 27, 1910. UA 36. Ser 04. Box 8 Folder 1, *Lick Observatory Records*, Special Collections, Mary Lea Shane Archives, University of California Santa Cruz.

terms was vital.[10] Campbell laid out the objectives of the research—how the expedition sought to gauge the movement of the solar system through space, trace the course of the stars, measure the distance between stars, study binary stars, and understand how the universe of stars had come into being. Campbell also described to Mills one of the groundbreaking discoveries that his father's patronage had enabled. The "Mills spectrograph in Mt. Hamilton and Chile," he explained, had "established the totally unexpected fact that one star certainly in every four or five, on the average is attended by a massive companion star, the two stars in each system revolving around each other in orbits which the spectrograph enables us to determine."[11]

The point was to emphasize how "intimately associated," on a global scale, these discoveries were with the Mills name.[12] Numerous publications and acknowledgements had resulted from Mills's donations, as well as gold medals from scientific associations around the world. For someone who had no experience in the world of scientists and barely understood astronomical research, perhaps the mention of gold medals and Parisian "royalty" could seem convincing. Campbell also described the paths that these discoveries opened, for they were "intimately connected" to the "origin of the stars, the history of the evolutionary processes which have brought them up to their condition today.... and to what the future has in

[10] "Within the last twenty years there has been developed the method of measuring the motions of the stars by means of a spectrograph attached to a powerful telescope. This has proved to be one of the most fruitful fields ever developed in astronomy." D.O. Mills Expedition, Expenses Santiago, May 3, 1910. UA 36. Ser 04. Box 12 Folder 3, *Lick Observatory Records*, Special Collections, Mary Lea Shane Archives, University of California Santa Cruz.

[11] D.O. Mills Expedition, Expenses Santiago, May 3, 1910. UA 36. Ser 04. Box 12 Folder 3, *Lick Observatory Records*, Special Collections, Mary Lea Shane Archives, University of California Santa Cruz.

[12] "Your father's name is intimately associated with these studies, in the literature of astronomy. To illustrate, Director Vogel of the great observatory at Potsdam, Germany, easily the first observatory in Europe, published in the Astrophysical Journal for July, 1900, page 386: 'Through the generosity of Mr. D. O. Mills, Professor Holden, at the time Director of the Lick Observatory, was able to have a spectrograph constructed for the Observatory (...) Professor Campbell has given an extended description of this instrument, which, mainly through his endeavors, has become the most noted instrument of its time (...), the 'Mills spectrograph' has in the last few years achieved surprising results.'" D.O. Mills Expedition, Expenses Santiago, May 3, 1910. UA 36. Ser 04. Box 12 Folder 3, *Lick Observatory Records*, Special Collections, Mary Lea Shane Archives, University of California Santa Cruz.

store for them."[13] The future of astronomical research, Campbell insisted, lay in observation from the southern hemisphere.

> There are dozens of large and active observatories in the northern hemisphere, but extremely few in the southern hemisphere. The Carnegie Institution of Washington is now expending $200,000 on another line of work in the southern hemisphere, with observing station located at San Luis, Argentina, to remove some of the arrears in just one special line of investigation. Prior to the year 1901 the Lick observatory and about eight other great northern observatories in the United States, England, France, Germany and Russia were engaged in spectrographic observations of stellar velocities...No observations of this kind were being secured in the southern hemisphere.[14]

Campbell then proposed the plans, which he had intended to discuss with the senior Mills, for a permanent location for the observatory. A donation of $250,000 would create an endowment to ensure the continuity of the research, which "would always be connected" with the Mills name. However, if this was not a possibility, perhaps Ogden could finance another extension of five or ten years for the Chile expedition.[15]

Campbell's long and detailed arguments were not enough to convince Mills to continue financing the expedition. Ogden responded that he did not find evidence that his father had intended to support the observatory's transformation into a permanent station. Nor was he inclined to extend the financing beyond the year, as his father had agreed.[16] Campbell insisted again on the benefits of investing in such

[13]D.O. Mills Expedition, Expenses Santiago, May 3, 1910. UA 36. Ser 04. Box 12 Folder 3, *Lick Observatory Records*, Special Collections, Mary Lea Shane Archives, University of California Santa Cruz.

[14]D.O. Mills Expedition, Expenses Santiago, May 3, 1910. UA 36. Ser 04. Box 12 Folder 3, *Lick Observatory Records*, Special Collections, Mary Lea Shane Archives, University of California Santa Cruz.

[15]D.O. Mills Expedition, Expenses Santiago, May 3, 1910. UA 36. Ser 04. Box 12 Folder 3, *Lick Observatory Records*, Special Collections, Mary Lea Shane Archives, University of California Santa Cruz.

[16]"I do not think it at all probable that any large endowment will be made by us (...) we probably should expect the work to be wound up at the end of this year, unless some very extraordinary result was promised from you and was in sight. The work already accomplished has been very satisfactory but I do not think my father had in view any permanent

an important scientific endeavor and of maintaining his father's legacy. He again requested financing for five or ten more years and promised to get letters from astronomers around the world who could attest to the investigation's crucial relevance.[17] Despite Mills's consistent refusals, Campbell managed to meet with him in late 1910 and talked him into extending more funds for the expedition. The money fell far short of the Campbell's hopes, but it would keep the observatory running for two more years.[18]

Campbell remained determined to win more time and, if possible, to make the project permanent. He continued to impress upon Mills the value of the project, and even tried to explain it in terms of mining: "To limit or to stop the development of my investigations in this line would seem to me just as the stopping of work in an extremely rich gold mine would seem to the mine owner."[19] He again brought the president Wheeler of the University of California into the negotiations. Naturally, he had to keep Wheeler abreast of the extension to July 1913 and he asked the president to send a thank you note to Mills, hoping perhaps that the prestige of the University might change Mills's mind.[20] Ogden, however, simply sent money as they had agreed and no longer responded to Campbell's insistent proposals.[21] Campbell then began to discuss the financing problem directly with Wheeler. Something must have worked because months later the astronomer then in charge of the expedition,

arrangement." D.O. Mills Expedition, Expenses Santiago, May 7, 1910. UA 36. Ser 04. Box 12 Folder 3, *Lick Observatory Records*, Special Collections, Mary Lea Shane Archives, University of California Santa Cruz.

[17]D.O. Mills Expedition, Expenses Santiago, May 9, 1910. UA 36. Ser 04. Box 12 Folder 3, *Lick Observatory Records*, Special Collections, Mary Lea Shane Archives, University of California Santa Cruz.

[18]D.O. Mills Expedition, Expenses Santiago, December 20, 1910. UA 36. Ser 04. Box 12 Folder 3, *Lick Observatory Records*, Special Collections, Mary Lea Shane Archives, University of California Santa Cruz.

[19]D.O. Mills Expedition, Expenses Santiago, December 20, 1910. UA 36. Ser 04. Box 12 Folder 3, *Lick Observatory Records*, Special Collections, Mary Lea Shane Archives, University of California Santa Cruz.

[20]D.O. Mills Expedition, Correspondence Mills Foundation 1905–1915, December 19, 1910. UA 36. Ser 04. Box 8 Folder 4, *Lick Observatory Records*, Special Collections, Mary Lea Shane Archives, University of California Santa Cruz.

[21]D.O. Mills Expedition, Expenses Santiago, December 29, 1910. UA 36. Ser 04. Box 12 Folder 3, *Lick Observatory Records*, Special Collections, Mary Lea Shane Archives, University of California Santa Cruz.

Joseph Moore, wrote Campbell with the news that the project had been extended for two more years, until July of 1915.[22]

5.2 Dealing with Lazy and Drunk Chilenos

Although by then the expedition had been in Santiago for several years, doing business and establishing agreements with the Chileans often led to headaches for the *gringos*. For example, Moore discovered that the rent for his house was considerably higher than that of his neighbors and he turned to his lawyer, Dr. Walker, who negotiated the rent down. But such haggling, Moore realized, often required making a scene. As he observed the negotiations between Dr. Walker and the manager, he heard the latter say that the rent had to be high because "the astronomer will have to pay the $150 for he can't move, he has to stay there on account of being close to the observatory." When they threatened to leave, the manager accepted a reduction to $135. "The part that chagrinned me," Moore wrote, "was to be able to understand the whole conversation and not to be able to say anything, for unless one speaks the language well he is at a disadvantage, the slightest hesitation for a word takes all the force out of what you say."[23]

Since the astronomers rotated out of Chile frequently, they did not learn to speak Chilean-style Spanish. Most only knew a couple of words when they arrived, and the Chilean particularities did not make learning the language any easier. This was important because the American researchers needed to make ties with locals, not just for establishing rental costs but also for varied manual tasks and services.

One crucial tie was the observatory's security service that the Santiago police had established upon the American's arrival. Nine years later, Moore wrote Campbell that, "our faithful old 'guardian' Juan Muñoz died at 2 o'clock this morning of pneumonia. Pneumonia appears to be very sudden and fatal among these people." He was sorrowful since the policeman was faithful and had been with the observatory since Wright's

[22]D.O. Mills Expedition, General Correspondence 1912–1919—J. Moore to W.W. Campbell, April 16, 1912, UA 36. Ser 04. Box 10 Folder 1, *Lick Observatory Records*, Special Collections, Mary Lea Shane Archives, University of California Santa Cruz.

[23]D.O. Mills Expedition, General Correspondence 1906–1911—J. Moore to W.W. Campbell, January 25, 1910, UA 36. Ser 04. Box 9 Folder 5, *Lick Observatory Records*, Special Collections, Mary Lea Shane Archives, University of California Santa Cruz.

time. "The ninth 'Commisaria' [sic] at once sent us another 'guardian' for they apparently knew from the first that Muñoz was fatally ill."[24]

Indeed, the subject of health was a concern for Moore whose wife suffered in the new climate and had only gotten over a worrisome cough after sleeping on the hill. Certainly, the astronomers' work entailed health risks that could not be easily treated, in their view, in Santiago. In 1912, Moore and his assistant Sanford experienced eye problems when they tried to adjust the mirror. "We were adjusting the collimation of the iron arc, and although we used some care in protecting our eyes with dark glasses, it appears not to have been sufficient. In the night my eyes became so badly inflamed that I came down leaving Mr. Sanford, whose eyes then showed no signs of trouble." Nonetheless, the next day Sanford had the same problem, which Moore attributed to effects of the ultra violet light.[25]

In addition to language barriers, the death of faithful servants, and health risks, the astronomers also worried about items that seemed to be missing in their shipments from California. Moore noted that he had received a thermostat from Mt. Hamilton, but a platinum wire was missing. "I am not sure," he wrote, "whether one was sent or whether it was lost (?) in the customhouse."[26]

Other obstacles, related mostly to the work habits and customs of the men who worked on occasional repairs at the observatory, also mounted. For the gringos, the Chileans were simply lazy drunkards. Moore expressed frustration and even anger. Due to repairs at their residence, Moore, his wife Frederica Chase, and Sanford had to camp on the hill.[27] Moore had assured his wife that the work would be finished soon; however, the workers failed to show up for a whole week. "Finally, the paper

[24] D.O. Mills Expedition, General Correspondence 1912–1919—J. Moore to W.W. Campbell, January 9, 1912, UA 36. Ser 04. Box 10 Folder 1, *Lick Observatory Records*, Special Collections, Mary Lea Shane Archives, University of California Santa Cruz.

[25] D.O. Mills Expedition, General Correspondence 1912–1919—J. Moore to W.W. Campbell, April 2, 1912, UA 36. Ser 04. Box 10 Folder 1, *Lick Observatory Records*, Special Collections, Mary Lea Shane Archives, University of California Santa Cruz.

[26] D.O. Mills Expedition, General Correspondence 1912–1919—J. Moore to W.W. Campbell, January 9, 1912, UA 36. Ser 04. Box 10 Folder 1, *Lick Observatory Records*, Special Collections, Mary Lea Shane Archives, University of California Santa Cruz.

[27] D.O. Mills Expedition, General Correspondence 1912–1919—J. Moore to W.W. Campbell, February 20, 1912, UA 36. Ser 04. Box 10 Folder 1, *Lick Observatory Records*, Special Collections, Mary Lea Shane Archives, University of California Santa Cruz.

hangers came bright and early and got everything already to work and then left word that they were going for 'desayuno' and would be back in a 'rato.' We did not see them again for three days."[28] When it came to fixing the floors, Moore decided to not wait for any Chilean workmen; rather, he and his wife waxed and polished the floors themselves.

Similar situations were recurrent. Months later, Moore himself, with the help of Sanford and his servant, had to take care of the electric lines. He had hired two "rotos," as working-class Chileans were pejoratively known, "but over noon they got drunk and after working an hour or so in the afternoon, threw down their tools and left without saying a word. We decided then that it was easier to do all of the work than to try and get some more of it out of these lazy Chilenos."[29] Labor problems were key to a report, regarding the possibility of living on the summit of San Cristóbal and therefore setting up a permanent station, that Campbell had requested of Moore. Living on the hill could reduce costs. Moore, of course, had been camping up there for almost a month in the summer. But he needed his servant to ride on horseback up the hill every morning with two jugs of water, milk, bread, and other provisions. The man often had to make two extenuating trips a day.

For Moore, living on the hill, even in a proper house, would mean that astronomers would have to travel to Chile alone without their families. Furthermore, the project would need service workers. "It might be possible," he wrote to Campbell, "to find a man and his wife, generally with a generous supply of children, who would be willing to live up there and the man be a mozo and his wife do the house work for the astronomers."[30] Still, this prospect would entail, in Moore's mind, several problems. Beyond the high birth rate among local families, he considered that the Chilean work habits made it impossible. He suggested hiring two servants. One "mozo" for heavy work and another one, less hearty but sober, who could work during the "four to eight days per month on

[28] D.O. Mills Expedition, General Correspondence 1912–1919—J. Moore to W.W. Campbell, January 9, 1912, UA 36. Ser 04. Box 10 Folder 1, *Lick Observatory Records*, Special Collections, Mary Lea Shane Archives, University of California Santa Cruz.

[29] D.O. Mills Expedition, General Correspondence 1912–1919—J. Moore to W.W. Campbell, December 10, 1912, UA 36. Ser 04. Box 10 Folder 1, *Lick Observatory Records*, Special Collections, Mary Lea Shane Archives, University of California Santa Cruz.

[30] D.O. Mills Expedition, General Correspondence 1912–1919—J. Moore to W.W. Campbell, February 20, 1912, UA 36. Ser 04. Box 10 Folder 1, *Lick Observatory Records*, Special Collections, Mary Lea Shane Archives, University of California Santa Cruz.

the average" when the first one was drunk.[31] Drunkenness was not the only problem that concerned Moore about Chilean workers—it rivaled with illnesses and illiteracy. He had to fire one worker who suffered from frequent stomach problems. A boy whom he hired later was a "willing worker" but "somewhat irresponsible." Furthermore, "he cannot read or write and even tell the time of day by clock."[32]

Despite these difficulties, the desire to work from the southern hemisphere remained strong. Some astronomers displayed an intense commitment that went beyond the prestige of science and their contractual obligations. The assistant Paddock, for example, decided to stay beyond the agreed time to continue his analysis of the plates that revealed important information about the orbit of binary stars. Even when Moore offered to pay him, Paddock volunteered to work "without other compensation than the privilege of doing his work here."[33] Paddock's eventual replacement, Mr. Sanford, adapted quickly to Santiago society and even went hunting for "huanacos" [sic] in the Andes that made for "quite a feast."[34] As Moore soon concluded, Sanford's work had been quite satisfactory, and he noted how popular he already was in the colony in Santiago. Perhaps Sanford liked being an American in Chile.[35]

As the weather reminded them, life in this far-away place was different from that in California. Snowfall in Santiago was an unusual event and its effects surprised the gringo astronomers. Moore wrote Campbell that,

[31] D.O. Mills Expedition, General Correspondence 1912–1919—J. Moore to W.W. Campbell, February 20, 1912, UA 36. Ser 04. Box 10 Folder 1, *Lick Observatory Records*, Special Collections, Mary Lea Shane Archives, University of California Santa Cruz.

[32] "The mozo which we have had since the middle of March can not read or write a word. I therefore made out the voucher for his salary and signed it myself." D.O. Mills Expedition, General Correspondence 1912–1919—J. Moore to W.W. Campbell, July 12, 1912, UA 36. Ser 04. Box 10 Folder 1, *Lick Observatory Records*, Special Collections, Mary Lea Shane Archives, University of California Santa Cruz.

[33] D.O. Mills Expedition, General Correspondence 1912–1919—J. Moore to W.W. Campbell, January 9, 1912, UA 36. Ser 04. Box 10 Folder 1, *Lick Observatory Records*, Special Collections, Mary Lea Shane Archives, University of California Santa Cruz.

[34] D.O. Mills Expedition, General Correspondence 1912–1919—J. Moore to W.W. Campbell, February 20, 1912, UA 36. Ser 04. Box 10 Folder 1, *Lick Observatory Records*, Special Collections, Mary Lea Shane Archives, University of California Santa Cruz.

[35] D.O. Mills Expedition, General Correspondence 1912–1919—J. Moore to W.W. Campbell, April 16, 1912, UA 36. Ser 04. Box 10 Folder 1, *Lick Observatory Records*, Special Collections, Mary Lea Shane Archives, University of California Santa Cruz.

"It was a rare sight indeed to see the orange trees full of ripe oranges and covered with snow. Many of the people in Santiago had never seen snow before except in the distant mountains. The snow does not seem to cause poor seeing here like it does at Mt. Hamilton."[36] The weather, to which Moore's wife never grew accustomed, was among the factors that made Moore wish to return in the summer of 1913.

> For more than a year, I have been feeling the effects of night work and at times pretty badly. Working all night is as you know, somewhat tiring, even when one is within a few steps of the observatory. When one is living at some distance from the observatory and at a difference of level of a thousand feet, night work seems to take all of the energy out of one, especially after three or four years of it.[37]

With financing assured for two more years, the project soon needed a replacement who could conduct research during the remaining time.

When the new astronomer Ralph E. Wilson arrived, he found similar challenges and problems. Relations with Chileans, be they *rotos* or fellow professionals, continued to be strained. Wilson attempted several times to meet with Ismael Gajardo, the subdirector of the Chilean National Astronomical Observatory, to discuss his work. After six or seven missed appointments, the two again agreed to meet, this time on the hill. However, when Gajardo sized up the ascent, he declined to make the walk up.[38]

Although the Chilean's behavior seemed strange, the Americans still needed to have some relationship with the nationally recognized astronomer. Indeed, engaging with Chileans was necessary for daily routines. Fortunately, the next head astronomer, Mr. Huffer, who arrived in August of 1917, was more enthusiastic about living in Chile. "I like Santiago very much and do not mind the climb up the hill to the

[36]D.O. Mills Expedition, General Correspondence 1912–1919—J. Moore to W.W. Campbell, July 12, 1912, UA 36. Ser 04. Box 10 Folder 1, *Lick Observatory Records*, Special Collections, Mary Lea Shane Archives, University of California Santa Cruz.

[37]D.O. Mills Expedition, General Correspondence 1912–1919—J. Moore to W.W. Campbell, October 29, 1912, UA 36. Ser 04. Box 10 Folder 1, *Lick Observatory Records*, Special Collections, Mary Lea Shane Archives, University of California Santa Cruz.

[38]D.O. Mills Expedition, General Correspondence 1912–1919—R. Wilson to W.W. Campbell, November 18, 1913, UA 36. Ser 04. Box 10 Folder 2, *Lick Observatory Records*, Special Collections, Mary Lea Shane Archives, University of California Santa Cruz.

observatory. I am very nicely located with a good Chilean family where the board is excellent."[39]

5.3 Making Science in Chile During the Great War

During the time between the tenures of Wilson and Huffer, the planet had of course changed markedly. The assassination of Franz Ferdinand set off a war that most supposed would last only a few months. Astronomers in Chile and the United States had close relations with Europe. The continent remained the "center of the earth;" everything related to culture, scientific progress, and the expansion of knowledge was rooted there. The importance of the United States was certainly growing but it was hardly clear then that the balance of world powers would shift so sharply. Since Europe dominated science and culture, trans-Atlantic trips were frequent and necessary. Campbell's own work had him traveling regularly to Europe, either to conduct research or to present his findings.

After Franz Ferdinand's death, the war spread throughout Europe. The previous period of "peace through strength," alliance policy, and growing nationalisms affected the development of what was to be known as the Great War. A place as remote as the southern tip of South America also became involved in the conflict in Europe. Wilson noted how Chileans felt connected to the war, despite the thousands of miles that separated them. "Santiago is very much excited over the news of general war in Europe... The streets are crowded with people reading the bulletins, business is at stand-still and the German banks have suspended payments entirely."[40] Although the Americans did not have banking problems in Chile, they had other, deeper concerns. Campbell and his colleagues were, precisely in the summer of 1914, on an expedition to observe an eclipse in Russia.[41] Wilson wondered if Campbell would be

[39] D.O. Mills Expedition, General Correspondence 1912–1919—Huffer to W.W. Campbell, August 28, 1917, UA 36. Ser 04. Box 10 Folder 6, *Lick Observatory Records*, Special Collections, Mary Lea Shane Archives, University of California Santa Cruz.

[40] D.O. Mills Expedition, General Correspondence 1912–1919—R. Wilson to W.W. Campbell, August 3, 1914, UA 36. Ser 04. Box 10 Folder 2, *Lick Observatory Records*, Special Collections, Mary Lea Shane Archives, University of California Santa Cruz.

[41] D.O. Mills Expedition, General Correspondence 1912–1919—R. Wilson to Tucker, August 3, 1914, UA 36. Ser 04. Box 10 Folder 2, *Lick Observatory Records*, Special Collections, Mary Lea Shane Archives, University of California Santa Cruz.

able to return to the United States. The war had by then fully broken out and trans-Atlantic voyages were all suspended.[42] Wilson marveled at how the war was affecting life in far-away Chile.

> Everything down here is war. Business is stagnated, no one knows what the exchange is, the streets are crowded with people watching the bulletins, the German banks are closed and many of the wholesale houses have stopped selling goods. There is quite a large percentage of Germans and French in Santiago and Valparaíso and we have had several incipient riots.[43]

Clearly, the forces of nationalism also arrived in the end of the world. Unrest and unemployment rose in Santiago and there were rumors of general labor strikes.[44] To keep fears of a revolution at bay, the Chilean government issued a massive amount of paper money and hired many of the unemployed.[45]

The astronomers faced several problems, especially the lack of materials and provisions necessary for their work. They had been purchasing their vacuum tubes, for example, from a German firm and now struggled to find other suppliers.[46] They resolved other supply problems in a more Latin American fashion. Although the government had prohibited the sale of distilled water, the astronomers found a way to convince a local laboratory to "make an exception."[47]

[42] "We are wondering if it will be possible to carry thru the eclipse expedition or if, with the regular trans-atlantic steamer service stopped, you will be able to get back to the States for some time." D.O. Mills Expedition, General Correspondence 1912–1919—R. Wilson to W.W. Campbell, August 3, 1914, UA 36. Ser 04. Box 10 Folder 2, *Lick Observatory Records*, Special Collections, Mary Lea Shane Archives, University of California Santa Cruz.

[43] D.O. Mills Expedition, General Correspondence 1912–1919—R. Wilson to Tucker, August 3, 1914, UA 36. Ser 04. Box 10 Folder 2, *Lick Observatory Records*, Special Collections, Mary Lea Shane Archives, University of California Santa Cruz.

[44] Beverly Hard, *Notas sobre los directores de la expedición de D.O. Mills a Chile*, w/d, code 08R-0101, Box 416, Archivo Histórico Universidad Católica de Chile.

[45] D.O. Mills Expedition, General Correspondence 1912–1919—R. Wilson to Tucker, September 1, 1914, UA 36. Ser 04. Box 10 Folder 2, *Lick Observatory Records*, Special Collections, Mary Lea Shane Archives, University of California Santa Cruz.

[46] D.O. Mills Expedition, Correspondence Mills Foundation 1905–1915, June 9, 1915, UA 36. Ser 04. Box 8 Folder 4, *Lick Observatory Records*, Special Collections, Mary Lea Shane Archives, University of California Santa Cruz.

[47] D.O. Mills Expedition, General Correspondence 1912–1919—R. Wilson to W.W. Campbell, August 27, 1914, UA 36. Ser 04. Box 10 Folder 6, *Lick Observatory Records*, Special Collections, Mary Lea Shane Archives, University of California Santa Cruz.

The war also caused other disruptions in far-away Chile. The astronomers had grown close to many of the Europeans who had settled in Santiago and the war caused significant tensions among them.

> While Americans are supposed to be neutral, the greater part of our friends are naturally among the English people. We have a good many German friends also, but, so far, we have been able to manage affairs so that the Germans and English have not met under our roof. Our German friends are quite rabid in their conversation and, while our English friends don't express their feelings so strongly, nevertheless the hatred of the Germans sticks out all over them.[48]

Wilson blamed the displacement of German scientists on international politics. Judging from Chileans' attitude toward all the Germans, he wrote, "I think that they would be very well satisfied to get rid of all of them." Wilson considered hiring some of the displaced Germans. They were willing to work hard with a telescope, but as "gentlemen-astronomers," they were "not very able or willing to do mechanical work."[49]

The rampant inflation in Santiago led Wilson to send his family home, sell his furniture, rent a nearby apartment, and resign his post suddenly in July 1918 without arranging for a replacement. His assistant, C. M. Huffer, a 23-year-old inexperienced student who was not even officially part of the Lick team, ran the station for over a year, during which he faced a series of calamities including influenza, vandalism, and faulty supplies. Finally, George Paddock, the former assistant, took over in 1919.[50] Managing the San Cristóbal station became even more challenging when the rural and uninhabited hill they encountered in 1903 underwent a transformation.

[48] D.O. Mills Expedition, General Correspondence 1912–1919—R. Wilson to W.W. Campbell, September 21, 1914, UA 36. Ser 04. Box 10 Folder 2, *Lick Observatory Records*, Special Collections, Mary Lea Shane Archives, University of California Santa Cruz.

[49] D.O. Mills Expedition, General Correspondence 1912–1919—R. Wilson to W.W. Campbell, September 21, 1914, UA 36. Ser 04. Box 10 Folder 2, *Lick Observatory Records*, Special Collections, Mary Lea Shane Archives, University of California Santa Cruz.

[50] Hard, *Notas sobre los directores.*

5.4 A Changing City: The Urban Park at San Cristóbal Hill

During the first years of 1900s, the city of Santiago had considerably transformed, and the American scientists witnessed those changes. Some of them affected their mission quite directly. Around the same time that the Mills astronomers were conducting their earliest observations on San Cristóbal Hill, in 1903 the archbishop of Santiago, Mariano Casanova, created a commission to discuss how to celebrate the 50th anniversary of the dogmatic definition of the Immaculate Conception that the church was to celebrate the following year. The commission decided to build a sanctuary in honor of the Virgin on the summit of San Cristóbal.[51] The hill was an icon in the history of Santiago. From the time of the conquistador Pedro de Valdivia, there was talk of a cross on the summit illuminated by candles left by the devout who hiked the hill.[52]

The church needed first to raise money, a task that fell to local society ladies eager to engage in religious and charitable endeavors that enhanced the cultural catch of the oligarchic *belle époque* capital city. In addition, the sanctuary was going to need land and in a place high enough that a monument to the Virgin would be visible throughout the city. At the commemoration of the anniversary of the Immaculate Conception dogma on December 8, 1904, Casanova blessed the first stone of the monument that would be finished four years later.

On the summit of San Cristóbal, science and religion, so often at odds, seemed to converge. Each benefitted from the other by, among other things, using the path and routes built to get to the top. In fact, the monument to the Virgin was crucial for the subsequent construction of the zigzagging road that ascends the hill from Bellavista neighborhood.[53] Science and religion, the Virgin Mary and the Observatory, should live together on the hill's twin summits. The local press highlighted the peculiar coexistence on the spot that for decades had been only the city's quarry. "Colored in the delicate shades of the afternoon, the hill stands full, showing the little paths serpent-like that go

[51] Juan Medina Torres, *El Cerro San Cristóbal. El gran balcón de Santiago* (Santiago: Consejo de Monumentos Nacionales, 2003), 19.

[52] *Libro Oficial del Cuarto Centenario de la Fundación de Santiago* (Santiago; Artuflo, 1941).

[53] *Revista En Viaje*, 181, November 1948, 2.

up its sides, the bloody wounds of the quarries, the grey bonnet the Observatory, and the Virgin, white and pure against the background of the sky, her arms reaching out toward immensity."[54]

On what Charles Darwin had described a century earlier as a "solitary hill that rises from the fog like an ancient island,"[55] science and religion coexisted in the monumentality of the observatory next to the Virgin. Indeed, as the press noted, both seemed to recognize the distance of space and time between them. "When the little red lights on the Virgin's diadem turn on like a lighthouse for the city's darkness, the colossal lens of the astronomers unfolds under the soft glow of the lights like a rigid antenna, searching in the immensity the cause of all mysteries."[56]

This celebration of San Cristóbal was remarkable, considering that most hills around Santiago had previously been deemed useful only for quarries.[57] The downtown Santa Lucía Hill was the exception. In the 1870s, the early-republican leader Benjamín Vicuña Mackenna had transformed Santa Lucía, bringing it into the space of the city.[58] Interestingly, the Gilliss mission had decided to install its observatory there in the mid-nineteenth century. Nonetheless, with the passage of time, those curious island-hills became part of the city as Santiago doubled in size between 1870 and 1915.[59] The city grew not only in size but, as the festivities for the centennial of independence showed, also in beauty.[60]

Calls to urbanize San Cristóbal, as well as the whole area north the Mapocho River, the imagined edge of Santiago populated by shacks and hovels, grew more prominent.[61] Shortly after the inauguration of

[54] "El San Cristóbal," *Pacífico Magazine*, November 1914, 600.

[55] Charles Darwin, *Viaje de un naturalista alrededor del mundo* (Buenos Aires: Ateneo, 1945), 310.

[56] "El San Cristóbal," 604.

[57] Carlos Martner, "El Cerro San Cristóbal en el paisaje de la ciudad. Obras desde 1960 y 1975," *Revista ARQ* 34 (December 1996): 5.

[58] Benjamin Vicuña Mackenna, *La transformación de Santiago: notas e indicaciones respetuosamente sometidas a la Ilustre Municipalidad, al Supremo Gobierno y al Congreso Nacional* (Santiago: Librería del Mercurio, 1872).

[59] Armando De Ramón, "Estudio de una Periferia Urbana. Santiago de Chile 1850–1900," *Historia* 20 (1985): 209.

[60] Bárbara Silva, *Identidad y nación. Patria Vieja, Centenario y Bicentenario* (Santiago: Lom, 2008), 73–106.

[61] René Martínez, *Santiago de Chile: los planos de su historia. Siglos XVI a XX: de aldea a metropolis* (Santiago: Universidad Central, 2007), 92.

the monument to the Virgin in 1909, the first campaign to reforest the hill was proposed. Though never carried out, the idea of making this and other hills into public parks was gaining strength.[62] In 1916, when Pedro Bannen and Alberto Mackenna drafted a proposal to expropriate the grounds of the hill, the idea began to take shape. It was the first step in the transformation that would become official the following year when a new law declared these hills to be public goods.[63]

When the American astronomers considered the possibility of making the expedition into a permanent station some years later, they also thought about purchasing the observatory grounds, as well as securing access to the roads.[64] The purchase of the land never went through. However, the changes taking shape on the hill required them to fortify their claims to the observatory grounds. In 1912, for example, the astronomers carried out the arduous task of bringing underground electricity cables and telephone lines and encasing them in wood to keep them out of the way of the construction of a new quarry elevator.[65]

Indeed, entanglements with the quarry became increasingly problematic for the astronomers. Wilson reported, "The old question of the graft off the observatory on account of the worthless stone-quarries has come again." He thought the affair was settled when Moore paid out the lawyers' fees. Nonetheless, he wrote, "the Chilean gentleman (?), Lecaros, was evidently expecting to be paid damages as well and I have received a letter from the lawyer, H. Walker Larrain, enclosing one from Lecaros asking 2,000 pesos as damages from 1911, when Dr. Moore stopped

[62] Beatriz Aguirre and Simon Castillo, *De la "Gran Aldea" a la ciudad de masas: El espacio público en Santiago de Chile, 1910–1929* (Santiago: Universidad Central, 2004), 18.

[63] "Ley num 3.295", *Diario Oficial*, September 28, 1917.

[64] "The lawyer, Mr. Walker told me that he thought we might possibly be able to buy the ground, should we ever care to (…) they couldn't sell it without giving the right of way (…) All of the holdings of foreign corporations (…) held in the name of Mission Boards in the States, and therefore in Chile coming under the head of 'Sociedad Anonyma,' have to get a special act of Congress in order to hold their property and this act must be renewed every five years, or the property is confiscated." D.O. Mills Expedition, General Correspondence 1912–1919—J. Moore to W.W. Campbell, February 20, 1912, UA 36. Ser 04. Box 10 Folder 1, *Lick Observatory Records*, Special Collections, Mary Lea Shane Archives, University of California Santa Cruz.

[65] D.O. Mills Expedition, General Correspondence 1912–1919—J. Moore to W.W. Campbell, December 10, 1912, UA 36. Ser 04. Box 10 Folder 1, *Lick Observatory Records*, Special Collections, Mary Lea Shane Archives, University of California Santa Cruz.

paying."[66] The Mills expedition was involved in some system of illegal payments developed by certain Chileans connected to the San Cristóbal quarries.

Wilson was aware of the absurdity of the bribery, but he also understood that "they will carry the bluff as far as to put men to work with dynamite as near as the observatory as possible and take out a little worthless stone."[67] Therefore, dynamite used in the quarry, from which little stone was even extracted by then, was a frequent subject in the correspondence between Chile and California. Years later, in 1917, Wilson wrote Campbell that gladly the work in the stone-quarry beneath the observatory stopped for some time.[68]

Yet, to the astronomers' chagrin, as soon as one problem was resolved another began. The astronomers no longer had to bribe the quarry to prevent its workers from detonating dynamite near the observatory; however, plans to transform the hill continued. The expropriation, announced in 1915, would soon become a reality.[69] The next year, the Boy Scouts carried out a symbolic "claiming" of the land that was soon followed by legal actions.[70] The astronomers considered that "the bill expropriating the cerro for a public park" would save them from "more trouble from the stone-quarry."[71] However, what the astronomers still did not know was that the landscape and the tranquility of the hill were going to change sharply. The following year, the intendant Pablo Urzúa took possession of the hill's grounds.[72]

[66] D.O. Mills Expedition, General Correspondence 1912–1919—R. Wilson to W. W. Campbell, November 18, 1913, UA 36. Ser 04. Box 10 Folder 2, *Lick Observatory Records*, Special Collections, Mary Lea Shane Archives, University of California Santa Cruz.

[67] D.O. Mills Expedition, General Correspondence 1912–1919—R. Wilson to W.W. Campbell, November 18, 1913, UA 36. Ser 04. Box 10 Folder 2, *Lick Observatory Records*, Special Collections, Mary Lea Shane Archives, University of California Santa Cruz.

[68] D.O. Mills Expedition, General Correspondence 1912–1919—R. Wilson to W.W. Campbell, August 27, 1917, UA 36. Ser 04. Box 10 Folder 6, *Lick Observatory Records*, Special Collections, Mary Lea Shane Archives, University of California Santa Cruz.

[69] *El Mercurio*, March 16, 1915.

[70] Medina, *El Cerro San Cristóbal*, 38ff.

[71] D.O. Mills Expedition, General Correspondence 1912–1919—R. Wilson to W.W. Campbell, August 27, 1917, UA 36. Ser 04. Box 10 Folder 6, *Lick Observatory Records*, Special Collections, Mary Lea Shane Archives, University of California Santa Cruz.

[72] Bárbara Silva and Josefina Cabrera, *Chile. Cien días en la historia del siglo XX* (Santiago: Planeta, 2015), 67.

From then on, the shape and role of Darwin's "island" in the city would change. It would no longer be a place for research far from the city noise. These changes brought all kinds of unforeseen problems, many related to the religious sanctuary. "Grass fires have repeatedly been set on slopes leading up to the observatory. It was necessary to burn off the grass around the enclosure and this a little later checked the flames on the fiesta of December 6."[73] The astronomers' need to find out what was in store for the hill became increasingly urgent. Chile's largest newspaper *El Mercurio* began publishing requests for projects such as restaurants and funiculars.[74] Clearly, the government sought to transform the hill into a public space meant for entertainment and sociability with easy access to the summit.

The city's incorporation of the hill became a reality some years later and San Cristóbal continued to grow as a place of entertainment for Santiago's residents. As Neubauer—the following acting astronomer in Chile- wrote, "The Cerro San Cristóbal ascensor is under construction and should be running next summer. There is also a line of motor buses running up on Sundays."[75] Such efficient transportation to the summit could be beneficial for the astronomers and the new restaurants could be not only functional but also even attractive during long stays. Nonetheless, these deep changes in the relationship between the hill and the city also altered the space where they conducted their work. As the local press noted, Santiago's intendant, Alberto Mackenna, was "determined to make the public his primary collaborator in the transformation of the San Cristóbal hill into one of the greatest parks in the world."[76]

[73] D.O. Mills Expedition, General Correspondence 1912–1919—G. Paddock to W.W. Campbell, December 29, 1919, UA 36. Ser 04. Box 10 Folder 8, *Lick Observatory Records*, Special Collections, Mary Lea Shane Archives, University of California Santa Cruz.

[74] *El Mercurio*, December 19, 1919. Paddock attached this clipping in his letter from December 29 of that same year, addressed to Campbell.

[75] D.O. Mills Expedition, General Correspondence 1920–1929—F. Neubauer to R. Aitken, May 14, 1924, UA 36. Ser 04. Box 11 Folder 2, *Lick Observatory Records*, Special Collections, Mary Lea Shane Archives, University of California Santa Cruz.

[76] Santiago Press clipping, no further identification, D.O. Mills Expedition, General Correspondence 1920–1929—F. Neubauer to R. Aitken, May 14, 1924, UA 36. Ser 04. Box 11 Folder 2, *Lick Observatory Records*, Special Collections, Mary Lea Shane Archives, University of California Santa Cruz.

Once the project to transform the hill gained momentum, it became necessary to decide what to do with the huge quarry or, as the press described it, the "enormous cavity that opens its jaws to the city."[77] The director of parks in Buenos Aires suggested the construction of terraces.[78] Other voices pressed for the conservation of the mountain's "rugged and solemn appearance" over a "coquettish decoration loaded with architecture."[79] The latter view coincided with the ongoing reforestation projects that inspired the motto "little architecture, much nature."[80] During 1924, changes on the hill were a constant concern for the astronomers and news of the San Cristóbal funicular even reached the American press.[81] Even though the astronomers could benefit from the easier access and more reliable electricity, all these changes would have consequences.

In 1923, when Campbell was appointed as President of the University of California, Robert G. Aitken was assigned as associate director of the Lick Observatory. Aitken tried to assuage Dr. Neubauer, then in charge of the observatory, and told him not to worry about the situation on the San Cristóbal, "Until the demands made by the authorities as to the manner in which we shall maintain

[77] Santiago press clipping, no further identification, D.O. Mills Expedition, General Correspondence 1920–1929 F.—Neubauer to R. Aitken, May 14, 1924, UA 36. Ser 04. Box 11 Folder 2, *Lick Observatory Records*, Special Collections, Mary Lea Shane Archives, University of California Santa Cruz.

[78] Medina, *El Cerro San Cristóbal*, 45

[79] Santiago Press clipping, no further identification, D.O. Mills Expedition, General Correspondence 1920–1929—F. Neubauer to R. Aitken, May 14, 1924, UA 36. Ser 04. Box 11 Folder 2, *Lick Observatory Records*, Special Collections, Mary Lea Shane Archives, University of California Santa Cruz.

[80] Santiago Press clipping, no further identification, D.O. Mills Expedition, General Correspondence 1920–1929—F. Neubauer to R. Aitken, May 14, 1924, UA 36. Ser 04. Box 11 Folder 2, *Lick Observatory Records*, Special Collections, Mary Lea Shane Archives, University of California Santa Cruz.

[81] "Construction work on the new tram-line to the top of San Cristóbal, Santiago. One of the largest transformers in South America, part of the General Electric Company's equipment recently installed in the San Cristóbal sub station, Santiago, 110,000 volt, 7,500 KVA" [Includes pictures from an American publication, apparently, from May 15, 1924], D.O. Mills Expedition, General Correspondence 1920–1929—F. Neubauer to R. Aitken, July 16, 1924, UA 36. Ser 04. Box 11 Folder 2, *Lick Observatory Records*, Special Collections, Mary Lea Shane Archives, University of California Santa Cruz.

our establishment becomes too exacting, we shall try to continue the Station in its present location."[82]

But by the mid 1920s, the hill had been undeniably transformed. The monument to the Virgin, the construction of restaurants, and then, the inauguration of the funicular and a zoo in 1925, radically changed San Cristóbal. Neubauer now had to accede to local officials' demands that the observatory replace its "unsightly" barbed-wired fence with a new fence that could keep away the "growing crowds" that gathered on the hill.[83] The once rugged place, frequented only by quarry workers, where the astronomers had found the silence and darkness that they needed for their observations of the stars, was gone forever. Added to administrative changes in Lick and the political and economic vicissitudes of the interwar years, the lights from restaurants and interruptions from the curious public who wanted to visit the observatory spelled the end of the project.

5.5 The University of California and Chile: Negotiations After 1920

Amid these changes in Santiago, shifts were also happening in California. The expedition, which had already lasted much longer than planned, had to find a new source of funds. In 1921, Campbell announced that the official title of the project was the Chile Station of the Lick Observatory, University of California. The reason for the change, as he explained, "is that moneys provided by Mr. Mills and his son…have been much more than consumed, and it is only fair that the reminder of the funds provided by others should not be made by the letterheads to seem to be contributed to the D.O. Mills expedition." In fact, he noted that "some men whom I approached for contributions objected to contributing to the support of the D.O. Mills Observatory."[84] A consequence of the

[82]D.O. Mills Expedition, General Correspondence 1920–1929, R. Aitken to F. Neubauer, August 25, 1924, UA 36. Ser 04. Box 11 Folder 2, *Lick Observatory Records*, Special Collections, Mary Lea Shane Archives, University of California Santa Cruz.

[83]Beverly Hard, "Notes on the directors of the D.O. Mills Expedition to Chile," 1972. Expeditions, UA 36. Ser 04. Box 12 Folder 6, *Lick Observatory Records*, Special Collections, Mary Lea Shane Archives, University of California Santa Cruz, 11.

[84]D.O. Mills Expedition, General Correspondence 1920–1929, W.W. Campbell to G. Paddock, January 26, 1921, UA 36. Ser 04. Box 11 Folder 1, *Lick Observatory Records*, Special Collections, Mary Lea Shane Archives, University of California Santa Cruz.

financial support of fourteen generous friends of the Lick Observatory, the name changes also highlighted the University of California's association with the scientific initiative in Chile.[85]

Although Campbell was able to solidify the expedition's standing, communication between him and the San Cristóbal staff faltered. He complained that Paddock had failed to inform him of a broken power line.[86] Then, in May 1922, Paddock abandoned his work in Chile without any notice to visit relatives in Rhode Island. Leaving his "personal assistant" in charge, he promised to return in July but did not reappear until mid-September. Soon after, when his tenure ended, Paddock purchased passage to Antofagasta on a steamship ten minutes before its departure and notified no one back at Mount Hamilton about when they might expect him.[87] Subsequently, Campbell named Neubauer to take over as director, a post he held until April 1929.

Several economic challenges greeted Neubauer's tenure. Much like Wright, Curtis, Moore, and Wilson, he complained to Campbell about the unexpectedly high cost of living and the difficult adjustment to life in a far-away country. He paid double the rent that Curtis had on a street that was considerably less elegant. Even the rent on his furniture increased.[88] "Once prices go up," he lamented, "they stay high." He had to live near San Cristóbal, "as no time should be added to the time required for climbing the hill." But since the university had not adjusted his salary to keep pace with rising costs, he could not live in a "good

[85] "Through the years 1917 to date the support of the institution has come from fourteen generous friends of the Lick Observatory. In 1919 the name D.O. Mills Expedition was changed to Chile Station of the Lick Observatory," William W. Campbell, "Notes from Pacific Coast Observatories," *Publications of the Astronomical Society of the Pacific* 40, no. 236 (1928): 251.

[86] D.O. Mills Expedition, General Correspondence 1920–1929—W.W. Campbell to G. Paddock, January 26, 1921, UA 36. Ser 04. Box 11 Folder 1, *Lick Observatory Records*, Special Collections, Mary Lea Shane Archives, University of California Santa Cruz.

[87] D.O. Mills Expedition, General Correspondence 1920–1929—F. Neubauer to R. Aitken, March 4, 1924, UA 36. Ser 04. Box 11 Folder 2, *Lick Observatory Records*, Special Collections, Mary Lea Shane Archives, University of California Santa Cruz.

[88] D.O. Mills Expedition, General Correspondence 1920–1929—F. Neubauer to R. Aitken, May 14, 1924, UA 36. Ser 04. Box 11 Folder 2, *Lick Observatory Records*, Special Collections, Mary Lea Shane Archives, University of California Santa Cruz.

section" of the city and instead settled near a sewer.[89] On top of all this, Neubauer told Aitken in California that the Chilean government had begun to tax the astronomers' salaries. This was no small matter. "It does not take much mathematics," as he put it, to understand his ambivalence about his future in the project.

The Chilean government had been urging middle-class families to buy houses to compel landlords to "tear down the old, unhealthy houses." Neubauer noted that most Americans and English residents were buying. Although the project lacked liquidity for any such purchases, Neubauer recommended this option if Campbell wished for him to stay for any length of time in this city that sought to be modern.[90] Urban renovation, as manifested in the incorporation of San Cristóbal Hill for example, also entailed the demolition of the city's old houses.

These challenges notwithstanding, Neubauer's life continued in the south of the planet. His family had accompanied him and remained through the difficult times, such as his daughter Phoebe Anne's bout with measles.[91] His relationships with local scientists and American organizations in Chile grew stronger. The Chilean Scientific Society, the American Club, and the YMCA all invited him to give talks. He also worked to remain in the good graces of the American diplomatic corps and, at the same time, resolve confusion that may have arisen from Paddock's hasty departure.

[89] "It isn't that I am at all dissatisfied with the work, or especially discontented with Chile (…) What I say about wishing or not wishing to stay after July 1925 boils itself down to the fact that we must live on a street where there is a sewer (…) and it does not take much mathematics to see that my salary does not go as far as the salaries paid in 1910 (…) I can not live in the good section of the city." D.O. Mills Expedition, General Correspondence 1920–1929—F. Neubauer to R. Aitken, May 14, 1924, UA 36. Ser 04. Box 11 Folder 2, *Lick Observatory Records*, Special Collections, Mary Lea Shane Archives, University of California Santa Cruz.

[90] D.O. Mills Expedition, General Correspondence 1920–1929—F. Neubauer to R. Aitken, May 14, 1924, UA 36. Ser 04. Box 11 Folder 2, *Lick Observatory Records*, Special Collections, Mary Lea Shane Archives, University of California Santa Cruz.

[91] "The winter is here now, very cold, and more so, our Phoebe Anne has been very sick, and is still in bed; it was a case of measles, with the characteristics of the Chilean measles, a heavy long fever." D.O. Mills Expedition, General Correspondence 1920–1929—F. Neubauer to R. Aitken, May 14, 1924, UA 36. Ser 04. Box 11 Folder 2, *Lick Observatory Records*, Special Collections, Mary Lea Shane Archives, University of California Santa Cruz.

> I met Mr. Collier, the American Ambassador, and he was under the impression that it was, the observatory, something supported by the Chilean government, or a branch of Harvard. The last I heard of Dr. Paddock was a postal from Rio dated April 16. So, by this time he will surely be at the Lick. I believe he had something to do with the new Harvard station in the north of Chile, as he went there, but of course he would never say a word. That is why Mr. Collier thought that we had something to do with Harvard.[92]

Maintaining a clear projection of the Lick Observatory's operation in Chile became especially important because a series of political events were quickly changing conditions in the country. In 1924, president Arturo Alessandri—known as the "Lion of Tarapacá"—faced widespread political convulsion. His administration struggled to carry out diverse social and political reforms but the parliamentary system that concentrated power in the legislature established virtually insurmountable obstacles. Proposed laws remained in congressional gridlock, making any decisions practically impossible.

The so-called "Sabre Rattling" occurred in September when young officers in the Chilean Army declared their discontent with the congress and demanded substantive reforms. It was the beginning of a deep crisis for Chile's institutions. In the short term, Alessandri withdrew from the country for six months; then, amid calls for his return, a new constitution was drafted. In 1925, the new laws reinstated a presidential system of government. Nonetheless, resolution of the crisis was still far away. In the following years, one of the Sabre Rattlers, Carlos Ibáñez del Campo, rose as a key figure in Chilean politics. In 1927, Ibáñez instituted a dictatorship that sought to modernize the country through a fortified state.

The resulting restrictions affected political and socioeconomic liberties in the country, as well as the work of the American astronomers. Chile was again orderly, stable, and in theory at least secure for investment in scientific research. However, the Chilean government no longer favored the observatory.[93] The political shifts imperiled the project. Despite its numerous successful spectrographic observations, it was increasingly

[92] D.O. Mills Expedition, General Correspondence 1920–1929—F. Neubauer to R. Aitken, May 14, 1924, UA 36. Ser 04. Box 11 Folder 2, *Lick Observatory Records*, Special Collections, Mary Lea Shane Archives, University of California Santa Cruz.

[93] Hard, "Notes on the Directors," 12.

difficult to argue that its labors were indispensable. In 1926, after the observatory's budget suffered a sharp decrease, Neubauer had to fire his assistant, Mr. Popp, and continue his work alone. Then, in July 1927, Aitken wrote that Campbell had decided to close the observatory in Chile as soon as Neubauer's contract ended.[94] The days of the Chilean station of the Lick observatory were numbered.

5.6 Farewell Begins

In addition to finalizing the ongoing observations, the astronomers launched into the negotiations regarding what to do with the observatory and all its equipment. One proposal was to sell everything to the Chilean government, if the price was right. Campbell approached the Chile's Consulate General in San Francisco but was soon disabused of the possibility. As Neubauer wrote Campbell, the chances that the Chilean government would sometime purchase the equipment were very slim, even at the price which was extremely low. Neubauer had already seen how the Chilean government had ordered a refractor for the National Observatory ten years earlier but had yet to pay for it. Furthermore, he noted, "The National Observatory is not interested in spectroscopy. The staff is composed of political appointments, and not one of the staff would stay as much as one hour after dark up on Cerro San Cristóbal."[95]

In the eyes of the Americans, politics intervened in the development of science in Chile, which was an obstacle to future collaboration in the country. But there were other factors that made selling the observatory difficult. Another was the "attitude of Chile toward anything 'yanqui,'" Neubauer wrote. "My first two years here were far from a bed of roses. And then when the Sacco and Vanzetti furor swept Chile, I was kept busy erasing threatening remarks from off the fence posts."[96] Beyond the international reaction that the Italian-Americans' death sentence

[94] Hard, "Notes on the Directors," 12.

[95] D.O. Mills Expedition, General Correspondence 1920–1929—F. Neubauer to W.W. Campbell, September 19, 1927, UA 36. Ser 04. Box 11 Folder 4, *Lick Observatory Records*, Special Collections, Mary Lea Shane Archives, University of California Santa Cruz.

[96] D.O. Mills Expedition, General Correspondence 1920–1929—F. Neubauer to W.W. Campbell, September 19, 1927, UA 36. Ser 04. Box 11 Folder 4, *Lick Observatory Records*, Special Collections, Mary Lea Shane Archives, University of California Santa Cruz.

provoked, Ibáñez's government took a sharply nationalist turn in mid-1927, which somehow translated into increased suspicion of the United States.[97]

Nonetheless, the astronomers remained confident that they could sell the observatory; they even had beautified the area at the summit. Neubauer considered this to be a positive factor that underscored his understanding of Chilean society in the 1920s. "As to the observatory itself," he wrote, "everything is in the best condition to make a favorable impression. The new paint, the fence, the rose vines and all—appearances are everything in Chile."[98]

He also recommended further steps, flowing from his long observation of local ways, to interest Chileans in the observatory. Neubauer suggested that Campbell play on the ego of Ismael Gajardo, director of the National Observatory, to persuade him to buy the equipment. "As I know the Chileans, I believe their sense of pride must be considered, and if you wrote to him directly about the matter, he would be pleased to receive a letter from 'DR. Campbell' that he might exert himself. He might publish it next day in the *Mercurio*, a custom here."[99] The cultural differences that Wright had perceived in the early years of his arrival in this strange country continued to surprise the American astronomers. More than two decades later, they still saw distances that hindered most spontaneous communication with Chileans.

Such asymmetries became even more evident as the departure loomed. Neubauer explained to Aitken, as he had already told Campbell, "The Chilean mind does not quite understand us and does interpret everything with a suspicion; and even if you gentlemen would give the observatory for a present, they still would want to be flattered

[97]This underlying suspicions toward the United States addressed economics and politics, notwithstanding the cultural influence of North America in Latin countries was indeed growing at accelerated pace. Stefan Rinke, *Encuentros con el Yanqui. Norteamericanización y cambio sociocultural en Chile.* 1898–1990 (Santiago: Dibam, 2013).

[98]D.O. Mills Expedition, General Correspondence 1920–1929—F. Neubauer to W.W. Campbell, September 19, 1927, UA 36. Ser 04. Box 11 Folder 4, *Lick Observatory Records*, Special Collections, Mary Lea Shane Archives, University of California Santa Cruz. Underlined in original.

[99]D.O. Mills Expedition, General Correspondence 1920–1929—F. Neubauer to W.W. Campbell, September 19, 1927, UA 36. Ser 04. Box 11 Folder 4, *Lick Observatory Records*, Special Collections, Mary Lea Shane Archives, University of California Santa Cruz.

in accepting it for nothing."[100] Therefore, he recommended that Aitken deal directly with the Chilean ambassador to the United States, taking note of how to conduct himself before the upper echelon of bureaucracy.

Although conversations with Gajardo could pave the way to the Secretariat of Public Education, which would make any decision, only higher authorities such as the ambassador could sway the government. Based on his hardened opinion of Chilean officials, Neubauer warned that the path would not be easy. "One point is certain. It will take a long time to close the bargain. It will take a longer time to get the money, and whatever will come do not let us figure the money to come for me to collect. Let that payment be made in the USA."[101] Neubauer did not make clear if he was referring to specific conditions, such as the recently installed Ibáñez government and the subsequent reorganization of the state that could hamper the transfer of funds, or to business transactions in Chile in general.

Another option that could circumvent the possible risks of dealing with the Chilean government involved establishing contact with the Catholic University of Chile. The Catholic University had been founded three decades earlier and was quickly growing.[102] The social characteristics of Chileans as described by Neubauer made this an attractive new possibility. "There is something in the air, which might make Dr. Campbell consent to sell the equipment here to the Catholic University."[103] The option revealed the web of relationships that, in Neubauer's view at least, was necessary to conducting business in Chile. He therefore described a Spanish priest, Luis Rodes, director of the Observatorio del Ebro, Spain, who could act as a bridge between the American sellers and the Chilean buyers from the university. Such

[100] D.O. Mills Expedition, General Correspondence 1920–1929—F. Neubauer to R. Aitken, September 22, 1927, UA 36. Ser 04. Box 11 Folder 4, *Lick Observatory Records*, Special Collections, Mary Lea Shane Archives, University of California Santa Cruz.

[101] D.O. Mills Expedition, General Correspondence 1920–1929—F. Neubauer to R. Aitken, September 22, 1927, UA 36. Ser 04. Box 11 Folder 4, *Lick Observatory Records*, Special Collections, Mary Lea Shane Archives, University of California Santa Cruz.

[102] Ricardo Krebs, *Historia de la Pontificia Universidad Católica de Chile 1888–1988* (Santiago: Ediciones UC, 1994), 219–224.

[103] D.O. Mills Expedition, General Correspondence 1920–1929—F. Neubauer to R. Aitken, October 11, 1927, UA 36. Ser 04. Box 11 Folder 4, *Lick Observatory Records*, Special Collections, Mary Lea Shane Archives, University of California Santa Cruz.

a Spaniard could favor the Chilean oligarchs' perception of American universities. Rodes, Neubauer wrote, "has been a guest of the Chilean government."

> He has been the 'lion' of society since the publication of his book 'El firmamento'...He is leaving today for California, and you will most likely meet him sometimes this month there. He gave three lectures in the university here which were a phenomenal success. It was the first time for me and Mrs. Neubauer in here we came listen with such joy as he certainly gave much credit to our American universities and observatories. He also gave a lecture in the Catholic University on American universities which was a sermon in itself.[104]

More to the point, the social relations that Rodes had developed could move along the plans for the observatory. "Rodes friends here, powerful Catholics of the '*más alta sociedad*' caught the germ from him and planning some means and way to put the Catholic University on the astronomical map. If they can raise the money they would like to take the observatory over and perhaps continue the work along our lines." The Catholic University, then, appeared to be a viable destination for the observatory and, in one way or another, keep the decades of effort from going to waste. "If Dr. Campbell prefers to make an offer directly," Neubauer wrote Aitken, "Perhaps he should write Monsignor Carlos Casanueva, rector of the Catholic University."[105]

Days later, in 1927, Neubauer insisted again on involving Father Rodes as a way to establish a more assertive link with university officials. Nonetheless, Neubauer pointed out that the university would undergo several changes regarding scientific research more broadly. Indeed, at the time, one of the nation's most important educational reforms was being structured in concert with Ibáñez's restructuring of government. That same year, the *Observatorio Nacional* was set as part of the Faculty of Physics and Mathematics of the State University.[106] Neubauer also

[104]D.O. Mills Expedition, General Correspondence 1920–1929—F. Neubauer to R. Aitken, October 11, 1927, UA 36. Ser 04. Box 11 Folder 4, *Lick Observatory Records*, Special Collections, Mary Lea Shane Archives, University of California Santa Cruz.

[105]D.O. Mills Expedition, General Correspondence 1920–1929—F. Neubauer to R. Aitken, October 11, 1927, UA 36. Ser 04. Box 11 Folder 4, *Lick Observatory Records*, Special Collections, Mary Lea Shane Archives, University of California Santa Cruz.

[106]Phillip Keenan, Sonia Pinto and Hector Álvarez, *The Chilean National Astronomical Observatory 1852–1965* (Santiago: Universidad de Chile, 1985), 145.

noted that the National Observatory lacked reliable communication, which perhaps confirmed his view that the government astronomers had little interest in astrophysics. Hence, he initiated contacts with Rector Casanueva and strove to convince him of the benefits of such a purchase.

> There is a possibility of purchasing the total equipment of our branch observatory here between May 1928 and May 1929. The equipment includes everything necessary for astrophysical investigations. The Price is more or less $8000 USA, hardly the fifth of the initial cost, certainly less than one tenth of the Price of what the observatory would cost if it were erected now, today. In the meanwhile, I am ready to have you visit the observatory at any time you may desire.[107]

Of course, he also invited the rector to visit the observatory. Then, in what was perhaps a gesture toward the Chileans social protocol that accompanied such negotiations, Campbell himself wrote Casanueva with an official offer. Campbell confirmed Neubauer's offer and signed off with the hope that "the offer I have made will interest you and be entirely satisfactory, I remain, with highest respect."[108] But Casanueva was already interested in buying the observatory. Even before receiving Campbell's letter, he had responded to Neubauer:

> The board of directors of our university has taken your proposal into consideration and has given it its approval in a general way, and they only hope that your university, that overflows in wealth and propriety shall bear in mind of fixing the price and conditions of their formal offer that they are helping the advance of scientific researcher and achievements through a comparatively young organization which owes its existence entirely to the efforts of private initiatives.[109]

[107] D.O. Mills Expedition, General Correspondence 1920–1929—F. Neubauer to C. Casanueva, October 18, 1927, UA 36. Ser 04. Box 11 Folder 4, *Lick Observatory Records*, Special Collections, Mary Lea Shane Archives, University of California Santa Cruz.

[108] D.O. Mills Expedition, General Correspondence 1920–1929—W.W. Campbell to C. Casanueva confirming Neubauer's letter, November 14, 1927, UA 36. Ser 04. Box 11 Folder 4, *Lick Observatory Records*, Special Collections, Mary Lea Shane Archives, University of California Santa Cruz.

[109] D.O. Mills Expedition, General Correspondence 1920–1929—C. Casanueva to F. Neubauer, November 9, 1927, UA 36. Ser 04. Box 11 Folder 4, *Lick Observatory Records*, Special Collections, Mary Lea Shane Archives, University of California Santa Cruz.

The Catholic University thus became directly involved with the American astronomers' project. The observatory would aid the university's goals to continue expanding, develop relationships that could contribute to its prestige, and create an international reputation. Casanueva's response underscores the drive to modernize the university by conducting state of the art research from the southern extreme of the world. All these factors intervened in the university's official recognition of Neubauer. "As a public testimony of your efficient services to the astronomical science in Chile," Casanueva informed Neubauer, "my Board has resolved to confer upon you the title of Member Honoris Causa of its Faculty of Physics and Mathematics."

In addition, the university invited Neubauer to give "a special course of astronomy as professor extraordinary."[110] The Honoris Causa recognition was a way to flatter and thank Neubauer and to publicize the association that the Catholic University was establishing, through Neubauer, with the Lick Observatory and the University of California. Indeed, Neubauer received his diploma from the hands of Manuel Foster. The San Cristóbal observatory eventually was named after Foster, who appeared as the official buyer of the American facility.[111]

The next steps included the gringo astronomers' definitive farewell. Neubauer remained in Chile a year after the observatory's sale in 1928. He accepted the offer to teach a course at the Catholic University and remained in charge of the observatory, in which he trained his successor, Rubén Toro Valenzuela, who held the post until 1939.[112]

The quarter century of history of the astronomers from California in Chile ended, though their legacy persisted through the Catholic University. They did not know it at the time, but decades later, other gringos would settle once again in the far-away latitudes of Chile. Meanwhile, as Neubauer said goodbye, he felt nostalgia for the place that had been his home for several years. The experience of living in the

[110] D.O. Mills Expedition, General Correspondence 1920–1929—C. Casanueva to F. Neubauer, November 9, 1927, UA 36. Ser 04. Box 11 Folder 4, *Lick Observatory Records*, Special Collections, Mary Lea Shane Archives, University of California Santa Cruz.

[111] "Reportaje Observatorio 'Manuel Foster.' Mirando a las estrellas," *Revista Noticias PUC*, n.d., D.O. Mills Expedition—History, UA 36. Ser 4. Box 12 Folder 6, *Lick Observatory Records*, Special Collections, Mary Lea Shane Archives, University of California Santa Cruz.

[112] "Reportaje Observatorio."

south—the distance, the surprises, the stars of another hemisphere—everything came together in a sensation of abandoning the place with which he had come to identify and where he had developed deep relationships. On July 4, 1929, he was back in California. Some 50 people saw him off at the Santiago train station. Neubauer "could not look toward Cerro San Cristóbal Hill for fear of crying like a child."[113]

That strange island-hill that rose in the middle of Santiago left him and the Lick Observatory behind. San Cristóbal was now part of Neubauer's past. Mount Hamilton received him anew and he would become accustomed to life in California. San Cristóbal would remain Santiago's still guardian, now an urban park. The hill continued to function as a window toward the stars, for several years at the service of Chile's scientific development.

Bibliography

Aguirre Beatriz and Simon Castillo, *De la "Gran Aldea" a la ciudad de masas: El espacio público en Santiago de Chile, 1910–1929*. Santiago: Universidad Central, 2004.

Campbell, William W. "Notes from Pacific Coast Observatories," *Publications of the Astronomical Society of the Pacific* 40, no. 236 (1928): 249–252.

Campbell, William W. *A Brief Account of the Lick Observatory of the University of California*. Berkeley: University of California Press, 1931.

Darwin, Charles. *Viaje de un naturalista alrededor del mundo*. Buenos Aires: Ateneo, 1945.

De Ramón, Armando. "Estudio de una Periferia Urbana. Santiago de Chile 1850–1900," *Historia* 20 (1985): 199–289.

D.O. Mills Expedition, UA 36. Ser 04. Boxes 1 to 12, *Lick Observatory Records*, Special Collections, Mary Lea Shane Archives, University of California Santa Cruz.

El Mercurio, March 16, 1915.

El Mercurio, December 19, 1919.

"El San Cristóbal," *Pacífico Magazine*, November 1914.

Hard, Beverly. *Notas sobre los directores de la expedición de D.O. Mills a Chile*, w/d, code 08R-0101, Box 416, Archivo Histórico Universidad Católica de Chile.

[113] Hard, "Notes on the Directors," 13.

Hard, Beverly. "Notes on the Directors of the D.O Mills Expedition to Chile," Expeditions, UA 36. Ser 04. Box 12 Folder 6, *Lick Observatory Records*, Special Collections, Mary Lea Shane Archives, University of California Santa Cruz, 1972.

Keenan, Phillip, Sonia Pinto, and Hector Álvarez, *The Chilean National Astronomical Observatory 1852–1965*. Santiago: Universidad de Chile, 1985.

Krebs, Ricardo. *Historia de la Pontificia Universidad Católica de Chile 1888–1988*. Santiago: Ediciones UC, 1994.

"Ley num 3.295," *Diario Oficial*, September 28, 1917.

Libro Oficial del Cuarto Centenario de la Fundación de Santiago. Santiago; Artuflo, 1941.

Martínez, René. *Santiago de Chile: los planos de su historia. Siglos XVI a XX: de aldea a metropolis*. Santiago: Universidad Central, 2007.

Martner, Carlos. "El Cerro San Cristóbal en el paisaje de la ciudad. Obras desde 1960 y 1975," *Revista ARQ* 34 (December 1996): 4–11.

Medina Torres, Juan. *El Cerro San Cristóbal. El gran balcón de Santiago*. Santiago: Consejo de Monumentos Nacionales, 2003.

"Reportaje Observatorio 'Manuel Foster.' Mirando a las estrellas," *Revista Noticias PUC*, D.O. Mills Expedition—History, UA 36. Ser 04. Box 12 Folder 6, *Lick Observatory Records*, Special Collections, Mary Lea Shane Archives, University of California Santa Cruz, n.d.

Revista En Viaje, 181, November 1948, 2.

Rinke, Stefan. *Encuentros con el Yanqui. Norteamericanización y cambio sociocultural en Chile*. 1898–1990. Santiago: Dibam, 2013.

Silva, Bárbara. *Identidad y nación. Patria Vieja, Centenario y Bicentenario*, 73–106. Santiago: Lom, 2008.

Silva, Bárbara and Josefina Cabrera. *Chile. Cien días en la historia del siglo XX*. Santiago: Planeta, 2015.

Vicuña Mackenna, Benjamin. *La transformación de Santiago: notas e indicaciones respetuosamente sometidas a la Ilustre Municipalidad, al Supremo Gobierno y al Congreso Nacional*. Santiago: Librería del Mercurio, 1872.

CHAPTER 6

Epilogue: Transnational Mirrors for the Stars

Abstract The Epilogue describes what happened with the observatory after the American scientists left, not imagining there was another chapter yet to be told for American Science in the south. Although there is no direct relationship, it connects this story to the further development in Astronomy, in the 1960s Chile, and sets the difference between how societies dealt with science in the early twentieth century, and then in the global sixties. Through the epilogue it is clear how this history of an observatory in the early twentieth century has much to say on the north-south approach. By analyzing this unique experience, and connecting it with the further development of astronomy in the sixties, we can open perspectives to understand the complexity of science and international relations.

Keywords Global sixties · Global Cold War · International astronomy in Chile

The American astronomers had planned to stay in Chile for two, maybe three years. In the end, they stayed for twenty-six—long enough for the

This section is based in the research of the FONDECYT project num. 3170099, PI Bárbara Silva, *Stars and Galaxies from the South: Chile and Its Scientific—Astronomical Insertion in the Global Cold War, 1962–1973.*

B. K. Silva, *Astronomy at the Turn of the Twentieth Century in Chile and the United States*, Palgrave Studies in the History of Science and Technology, https://doi.org/10.1007/978-3-030-17712-6_6

expedition to see changes in the scientists, astronomy, and the host society. Acting astronomers rotated out every three to four years and their return to California was often difficult. When the last astronomer in the southern country, Dr. Neubauer, boarded the train that took him to Valparaíso, the port of his departure to the United States, he could not look back at San Cristóbal. Emotions must have been everywhere for him: distance, different cultures, impressive mountains, annoying habits, the peaceful city, welcoming attitudes. Many sensations confused his feelings upon leaving Chile. The summit of San Cristóbal was a home for many American scientists during those years. Going up and down, they forged a meaningful relationship with Chileans and the faraway country. For them, that part of their lives was about to end, and it was indeed sad.

What they did not know was that a new chapter on American astronomy in Chile was about to be written. When the astronomers from California left Chile, the Catholic University took over the facilities. Although there were no Chilean astronomers to take this responsibility, Neubauer's former assistant, Ruben Toro, assumed the post of substitute director. Afterward, Erich Heilmaier, a German astronomer, was appointed as the Director of The Manuel Foster Observatory—the observatory's new name, after the Chilean man who bought it to donate it to the University. Heilmaier kept the observatory running, despite several difficulties, including lack of astronomers and lack of funding. In addition, there was no running water, the constructions needed several renovations, and then a huge fire damaged part of the observatory. Still he managed to overcome these drawbacks and conducted important astronomical research, most notably regarding the Star β Crucis.[1]

By the late 1940s, the situation was critical. Heilmaier criticized the lack of basic conditions for astronomical research. Then, in a major accident, Assistant Toro lost an eye.[2] Although the observatory continued to enhance its outreach mission, always as the property of the Catholic University, officials suspended research. For years, astronomers limited their work to measuring the huge amount of plates they already had.

[1] Erich Heilmaier, "Informaciones preliminares sobre la velocidad radial de β Crucis," *Revista Universitaria* 23, no. 2 (1938): 187.

[2] Andreas Reisenegger, "Estrella del pasado," *Revista Universitaria* 116 (2012): 25.

Then, in the late 1950s, amid the Cold War, astronomy in Chile revealed a new face. The Cold War promoted cutting-edge technological development, the exploratory phases of which were usually related and financed by the military.[3] Besides the scientific progress itself, technology was a very important dimension through which the world measured the superpowers' hegemony. Therefore, science and technology were relevant not only in military terms, but also for political and cultural influence, interwoven with their global projects of supremacy.[4]

In the critical years of the ideological dispute, the United States and the Soviet Union revealed a power struggle that materialized in different dimensions. The resulting space race produced a particular sensation of novelty and uncertainty. The global confrontation expanded into a multilateral and extra-atmospheric arena that brought together science, politics, military power, and ideologies.[5] The 1957 Sputnik, the creation of NASA, the Apollo missions of the 1960s and 1970s, and even their expression in sci-fi literature and movies are examples of the space race's pervasive relevance.[6] The competition was taking place in space and learning more about the universe was vital. Though not directly related to aerospace knowledge, by straining the borders of the unfathomable, astronomy had an important role in the equation. Astronomy sought to understand what is going on "up there," where our eyes could no longer see. Now, the helpless feeling of looking up to the skies became a dimension of power.

[3] David Reynolds, "Science, Technology, and the Cold War," *The Cambridge History of the Cold War*, Vol. 3, eds. Melvyn P. Leffler and Odd Arne Westad (Cambridge: Cambridge University Press, 2010), 399.

[4] Nils Gilman, *Mandarins of the Future: Modernization Theory in the Cold War America* (Baltimore: The Johns Hopkins University Press, 2007), 7.

[5] John Krige, "Embedding the National in the Global: US–French Relationships in Space Science and Rocketry in the 1960s," *Science and Technology in the Global Cold War*, eds. Naomi Oreskes and John Krige (Cambridge: Massachusetts Institute of Technology Press, 2014), 230.

[6] George Reisch, "When Structure Met Sputnik: On the Cold War Origins of the Structure of Scientific Revolutions," *Science and Technology in the Global Cold War*, eds. Naomi Oreskes and John Krige (Cambridge: Massachusetts Institute of Technology Press, 2014), 371; Erik M. Conway, "Bringing NASA Back to Earth: A Search for Relevance During the Cold War," *Science and Technology in the Global Cold War*, eds. Naomi Oreskes and John Krige (Cambridge: Massachusetts Institute of Technology Press, 2014), 251.

In the 1950s, it was clear the southern skies offered rich potential for astronomical research,[7] as the Lick astronomers and other scientific missions in the late nineteenth and early twentieth centuries proved. Magellanic Clouds, for example, delivered important information only accessible from the southern hemisphere. As technology evolved, astronomers looked for places with less humidity and light pollution. In the late 1950s, the Association of Universities for Research in Astronomy (AURA), an American organization, began searching the southern hemisphere for the best site for an astronomical observatory. The search led scientists to uninhabited semiarid zones in Australia, South Africa, and South America.[8] At first sight, Australia or South Africa could have some leverage, since an English speaking community could facilitate administrative work.

But there were also political reasons to consider. On the one hand, the relationship that Australia and South Africa had with the British government made agreements difficult; Americans would have to deal with the United Kingdom, and also with a Third World country.[9] On the other hand, they had to be very careful, since hi-tech investments were extremely expensive. Scientists could not risk losing sophisticated equipment amid unstable political conditions.

These concerns partially explain the need to look in places such as South America. In the previous relationships between Chilean and American observatories, the southern country acted mostly as a host and the development of Chilean astronomy was extremely limited. But this was to drastically change in the 1960s.[10] Federico Rutllant, then the director of the National Astronomical Observatory of Chile, established contact with Gerard Kuiper, from the University of Chicago, and later with AURA. This relationship was decisive for the exploration of astronomical development in Chile.[11]

[7]"Convention Portant la Création d'une Organisation Européenne pour des Recherches Astronomiques dans l'Hémisphère Austral, signée à Paris, 5 Octobre 1962." *Archivo del Ministerio de Relaciones Exteriores de Chile*, Fondo Organismos Internacionales, Vol. 400.

[8]Hilmar W. Duerbeck, "National and International Activities in Chile 1849–2002," *Astronomical Society of the Pacific* 292 (2003): 9.

[9]"Letter J.W. Joyce, Acting Deputy to Paul A. Siple, "Scientific Attaché, American Embassy Canberra," February 28, 1966, Record Group 59, Entry 3008D, Box 20 Folder SCI 21 Visits, Missions, Astronomy, *National Archives College Park*, USA.

[10]Hernán Quintana and Augusto Salinas, "Cuatro siglos de astronomía en Chile," *Revista Universitaria* 83 (2004): 55.

[11]Frank K. Edmonson, *AURA and Its US National Observatories* (New York: Cambridge University Press, 1997), 137.

This led to the decision of American Scientists to carry out site testing in the southern country. Jürgen Stock, a German astronomer working for AURA explored the semiarid area of Chile from April to December 1959.[12] Through agreements with the Chilean government and the partial funding of the National Science Foundation, AURA developed a plan to install a large-scale observatory in Chile. In 1962, AURA decided to build the Cerro Tololo Interamerican Observatory in the Coquimbo region.[13]

The antecedent of the Lick Observatory in Chile had some influence in this process. One physics scholar notes, that "The experience of the Lick was decisive when the planning of the great American and European Observatories in the Southern Hemisphere began in the 1950s. Although Curtis' reports slept in the archives, the Lick astronomers knew of the excellent observation conditions in Chile."[14] Even if the influence of the Curtis report is questionable, it is still reasonable to suppose that Curtis taught the American scientists much about the area. More than five decades later, the habitability of the desert had certainly changed. Exploration of the Chilean deserts again seemed like a viable endeavor.

Other international initiatives involved the same area of Chile. The recently created European Southern Observatory (ESO) was looking for places to settle in the southern hemisphere as well. At first ESO showed little interest in Chile, but soon their plans changed. One year later, ESO officials said they would have to review the possibility of a radical change regarding the location of their observatory, due to favorable results of the American site testing in Chile.[15] Different reasons made them take separate paths, but Europeans and Americans remained in Chile. Both started plans of building massive observatories, one on *La Silla* and the other on *Tololo*. A few years later,

[12] Jürgen Stock, "Chile Observatory Project: Seeing Expedition," *Stock Reports*, w/d, Cerro Tololo Interamerican Observatory Library, Coquimbo, Chile.

[13] *El Mercurio*, November 24, 1962, 25.

[14] Ricardo Leiva, "Atacama: a 100 años del informe Curtis," *Revista de Humanidades* 3 (2010): 21.

[15] J.H. Oort, "Documents Pertaining to the Secretariat of the ESO Committee," quoted in Adriaan Blaauw, *ESO's Early History: The European Southern Observatory from Concept to Reality* (München: ESO, 1991), 48.

Carnegie Institution landed in the country and built its own observatory in the same semiarid area of Coquimbo. Not long after that, ESO projected a newer and bigger observatory in *Paranal*, approximately 800 kilometers north of Coquimbo in the middle of the Atacama Desert.

To make the scenario even more complex, in those same years, astronomers from the Soviet Pulkovo Observatory were collaborating with scientists working at the National Observatory. The Soviets' work in Chile expanded from initial collaborative project to building their own observatory with a Maksutov astrograph on Cerro Robles in 1967—one of the first sites Jürgen Stock tested for the AURA mission.[16] The relationship between the Soviet, American, and European astronomers in Chile was complicated. These were the Cold War years and it was not by chance they all ended up in the same place in faraway Chile. As AURA Director Edmonson recalled, "several members of the AURA Board expressed concern about the Soviet presence in Cerro Calan" at the National Observatory of Chile.[17]

Nonetheless, the presence of multiple international scientists in the country allowed Chileans to engage with scientific dimensions in ways that otherwise would have been impossible. Despite having been hosts for at least two different American missions in astronomy, their own development in the field was still limited. Accordingly, they signed agreements with all parties involved: negotiations with Americans and Europeans implied the presence of the University of Chile and also the diplomatic corps and the Ministry of Foreign Affairs. The purpose was to secure investments; in return, the government would provide full diplomatic status and exemption from taxes and customs fees.[18] These negotiations, begun under the right-wing government of Alessandri, were ratified under the Christian Democrat government of Eduardo Frei. Both presidencies had a close relationship with the United States

[16] Duerbeck, "National and International Activities," 18.

[17] Edmonson, *AURA and Its US National Observatories*, 202.

[18] "Carta de Luis Cubillos, Ministerio de Relaciones Exteriores a Otto Heckmann," November 27, 1963, Fondo Organismos Internacionales, Varios, Vol. 308, file 366. *Archivo del Ministerio de Relaciones Exteriores de Chile.*

through the Alliance for Progress.[19] With an anticommunist government in power, agreements with Pulkovo and the Soviet Academy of Sciences were restricted to the University of Chile.

Overall, the relationship between Americans and the Chilean scientific community was close, unlike the one with ESO. Authorities at AURA stated: "Chileans will have the opportunity to use the telescope and even if in Chile there are not many astronomers, the equipment will be loaned for a useful association in this sense."[20] The Carnegie Institution offered scholarships that sought to "contribute, in Chile, to the progress of this science, to improve scientific education, and to the professional development of young astronomers."[21] The Chilean-American cooperation included working with the novel Chilean community of astronomers, an agreement that was not reached with the Europeans during those years.

In only a few years, astronomy in Chile radically changed. At first, Chileans played their traditional role of hosts, but soon after the country enhanced its astronomical potential and strove to produce local science. The drive came from international interests settled in the middle of the Cold War. The converging factors of geography, politics, ideologies, and science led to a key moment for both Chile and astronomical research. The country's previous history in astronomy, and their earlier relationship with American science—especially the 1903–1929 Lick adventure—might have influenced these outcomes.

In the meantime, the southern station, renamed the Manuel Foster Observatory, remained on San Cristóbal. It was closed for research from the late 1940s to 1980. The dictatorship of Augusto Pinochet took over universities after the coup in 1973 and designated military rectors. Jorge Sweet was appointed the Catholic University rector and officiated the

[19]The Alliance for Progress was Kennedy's strategy to contain the spread of communism in Latin America, after the Cuban Revolution. It was officially announced in the 1961 Punta del Este Conference, as a collaborative plan between the US government and Latin American states. It sought strategic reforms for the region, such as Agrarian Reform, Educational Reform, housing, trading modernization, among others.

[20]"Apoyo a la investigación astronómica," *El Mercurio*, May 6, 1967, 1.

[21]"Panorama. Beca para astrónomos," *Las Últimas Noticias*, March 20, 1969, 6.

observatory's reopening ceremony in 1980. From the 1960s onwards astronomy had significantly developed in the southern country. The military government expelled the Soviets immediately, but the Americans and the Europeans continued to invest in astronomical research.

Although the San Cristóbal Observatory had long ceased to be a cutting-edge facility, it did allow the university to establish collaboration agreements with international observatories. The reopening also aimed to train students. However, this effort would be short. In 1995, the observatory closed the dome for good. Lights, smog, and nearby telephone antennas all made the observatory obsolete. Technology had evolved and research at San Cristóbal was not worth the investment to be made to keep it running. For almost a century, the San Cristóbal Observatory provided plates for radial velocities measurements. From the summit, it was a silent witness to all the changes in Santiago. It was the place for the California scientists' adventures and discoveries, but it was also a new challenge for a local university starting in the 1930s. It was where several fell in love with studying the universe.

William Wallace Campbell never thought his lifetime dream of having a southern station would have lasted so long. Nor could he have suspected that the southern country he picked would become an astronomical center. The 1990s showed how massive projects of new observatories developed in Chile.[22] These projects connected the experiences of the 1960s with the technology of the approaching twenty-first century. The exponential growth of astronomy in Chile, decisively sustained by international initiatives, keeps growing. These mega-investments took shape through projects such as VLT (Very Large Telescope) at Paranal, ALMA (Atacama Large Millimeter Array), and carries on.[23] Throughout, Chilean science has evolved. Slowly, there was a hint of consciousness of astronomy in the southern country.

In 2010, after years of being closed, the San Cristóbal observatory tried to take on a new life as a historical monument.[24] A new story was about to start for the centennial observatory—not for

[22] Sebastián López, et al. "Astronomía," *Análisis y Proyecciones de la Ciencia Chilena* (Santiago: CONICYT, 2005), 239–259.

[23] Javiera Barandiaran, "Reaching for the Stars? Astronomy and Growth in Chile," *Minerva* 53 (2015): 141–164.

[24] Consejo de Monumentos Nacionales, *Decreto 0352*, September 3, 2010.

astronomical research but for heritage value. With this category, the University sought to protect the building and its history, opening its doors once a year for Heritage Day.[25] Although renovations have not been done, and there is no visitors' center or scientific outreach facilities, citizens hike up to the summit to see this almost-forgotten historical place for astronomy. The heritage value converges with a wider trend to encourage appreciation of astronomy. UNESCO has promoted "Astronomy and World Heritage" to try to make astronomical heritage more visible.[26] Efforts include discussions about material and immaterial heritage categories, and how to conserve the darkness of night skies and therefore the visibility of stars.[27]

The San Cristóbal hill is now in the middle of Santiago. Toward the second summit, there is an abandoned dance floor. Beyond the dusty black and white squares, there is a dirt road covered by foliage. At first sight, it seems like a dead end. It is hard to imagine that once upon a time the hill, then a quarry, had no vegetation. Following the mysterious road, behind the trees and bushes, the strange round building remains. Below massive telecommunication antennas, the dome that once hosted the biggest spectrograph of the southern hemisphere shyly shines. The telescope still has the mirrors that reflected life in Chile and in California at the beginning of the twentieth century. Up the soft slope that takes the hiker to the second summit, the observatory almost stops time, waiting to be reopened and host, if not the cutting-edge science it once produced, at least a place for scientific outreach. Years have passed and the scenario the California astronomers found in Chile is gone. One can barely imagine this history in the hidden corners of the hill and the city. The observatory is there to remind us that, despite time, we still look up to the stars, and seek to understand lights in the darkness that contains the mysteries of the universe. It reminds us we are still able to find beauty in the unknown.

[25] "Telescopio del San Cristóbal inaugurado hace mas de 100 años vuelve a abrir sus puertas," *La Tercera*, May 22, 2012.

[26] Anna Sidorenko, "The UNESCO Thematic Initiative 'Astronomy and World Heritage'," *Astronomy in Focus*, Vol. 1, XXIXth IAU General Assembly (August 2015): 83–88.

[27] Michel Cotte, "What Makes Astronomical Heritage Vaulable? Identifting Potentias Outstanding Universal Value in Cultural Properties Relating to Astronomy," *Astronomy in Focus*, Vol. 1, XXIXth IAU General Assembly (August 2015): 93–96.

Bibliography

"Apoyo a la investigación astronómica," *El Mercurio*, May 6, 1967.

Barandiaran, Javiera. "Reaching for the Stars? Astronomy and Growth in Chile," *Minerva* 53 (2015): 141–164.

Blaauw, Adriaan. *ESO's Early History: The European Southern Observatory from Concept to Reality*. München: ESO, 1991.

"Carta de Luis Cubillos, Ministerio de Relaciones Exteriores a Otto Heckmann," *Archivo del Ministerio de Relaciones Exteriores de Chile*, November 27, 1963, Fondo Organismos Internacionales, Varios, Vol. 308, file 366. *Archivo del Ministerio de Relaciones Exteriores de Chile.*

Consejo de Monumentos Nacionales, *Decreto 0352,* September 3, 2010.

"Convention Portant la Création d'une Organisation Européenne pour des Recherches Astronomiques dans l'Hémisphère Austral, signée à Paris, 5 Octobre 1962," *Archivo del Ministerio de Relaciones Exteriores de Chile*, Fondo Organismos Internacionales, Varios, Vol. 400.

Cotte, Michel. "What Makes Astronomical Heritage Valuable? Identifying Potentials Outstanding Universal Value in Cultural Properties Relating to Astronomy," *Astronomy in Focus*, Vol. 1, XXIXth IAU General Assembly (August 2015): 93–96.

D.O. Mills Expedition, UA 36. Ser 04. Boxes 1 to 12, *Lick Observatory Records*, Special Collections, Mary Lea Shane Archives, University of California Santa Cruz.

Duerbeck, Hilmar W. "National and International Astronomical Activities in Chile," *Astronomical Society of the Pacific Conference Series* 292 (2003): 3–20.

Edmonson, Frank K. *AURA and Its US National Observatories*. New York: Cambridge University Press, 1997.

El Mercurio, November 24, 1962, 25.

Gilman, Nils *Mandarins of the Future: Modernization Theory in the Cold War America*. Baltimore: The Johns Hopkins University Press, 2007.

Heilmaier, Erich. "Informaciones preliminares sobre la velocidad radial de β Crucis," *Revista Universitaria* 23, no 2 (1938): 180–198.

Leiva, Ricardo. "Atacama: a 100 años del informe Curtis," *Revista de Humanidades* 3 (2010): 10–22.

"Letter J. W. Joyce, Acting Deputy to Paul A: Siple, Scientific Attaché, American Embassy Canberra," February 28, 1966, Record Group 59, Entry 3008D, Box 20 Folder SCI 21 Visits, Missions, Astronomy, *National Archives College Park*, USA.

López, Sebastián, Mario Hamuy, Luis Felipe Barrientos, José Maza, and María Teresa Ruiz. "Astronomía," *Análisis y Proyecciones de la Ciencia Chilena,* 239–260. Santiago: CONICYT, 2005.

Oreskes, Naomi and John Krige, eds. *Science and Technology in the Global Cold War*. Cambridge: Massachusetts Institute of Technology Press, 2014.

"Panorama. Beca para astrónomos," *Las Últimas Noticias*, March 20, 1969.

Quintana, Hernán and Augusto Salinas. "Cuatro siglos de astronomía en Chile," *Revista Universitaria* 83 (2004): 53–60.

Reisenegger, Andreas. "Estrella del pasado", *Revista Universitaria* 116 (2012): 24–27.

Reynolds, David. "Science, Technology, and the Cold War," *The Cambridge History of the Cold War*, Vol. 3, eds. Melvyn P. Leffler and Odd Arne Westad, 378–399 Cambridge: Cambridge University Press, 2010.

Sidorenko, Anna. "The UNESCO Thematic Initiative 'Astronomy and World Heritage'," *Astronomy in Focus*, Vol. 1, XXIXth IAU General Assembly (August 2015): 83–88.

Stock, Jürgen. "Chile Observatory Project: Seeing Expedition." *Stock Reports*, Cerro Tololo Interamerican Observatory Library, Coquimbo, Chile, n.d.

"Telescopio del San Cristóbal inaugurado hace mas de 100 años vuelve a abrir sus puertas," *La Tercera*, May 22, 2012.

Index

B. K. Silva, *Astronomy at the Turn of the Twentieth Century in Chile and the United States*, Palgrave Studies in the History of Science and Technology, https://doi.org/10.1007/978-3-030-17712-6

The manufacturer's authorised representative in the EU is Springer Nature Customer Service Centre GmbH, Europaplatz 3, 69115 Heidelberg, Germany. If you have any concerns regarding our products, please contact ProductSafety@springernature.com

Printed and bound by CPI Group (UK) Ltd, Croydon, CR0 4YY

30/09/2026

02225260-0005